Bringing Government into the 21st Century

ISBN (paper): 978-1-4648-0881-4
ISBN (electronic): 978-1-4648-0882-1
DOI: 10.1596/978-1-4648-0881-4

Cover photo: KDI School. Used with permission. Further permission required for reuse.
Cover design: Debra Naylor, Naylor Design, Inc.

Library of Congress Cataloging-in-Publication Data has been requested

Contents

Boxes

Figures

Tables

Foreword

The 21st century is the digital age. The challenge facing us as individuals is how to make the most of the new technology that is now available. We talk to friends who seem more knowledgeable than ourselves, and learn from their experience. Yet our efforts often leave us frustrated; we may have invested in new equipment that soon becomes outdated or perhaps we do not have the necessary skills. Overall, however, we are moving forward, not always along the most direct path, but learning from our mistakes and acquiring the skills we need in the 21st century. The same approach is true of Korea, as it brought government into the 21st century with such success that it is now the highest ranked country in the world for e-governance.

But why is Korea's decades-long journey so intriguing?

I think it's because of the flexibility, perseverance, and commitment, as well as the culture of pursuing results and dealing with any setbacks with renewed vigor. In adopting a digital approach to governance, the outcomes, changes and competencies expected of decision-makers and implementers are never easily achieved. While no recipe book can cover all conditions and contexts, this book provides a range of invaluable insights.

Implementing digital governance is challenging. Doing it well, persistently and continuously adapting to changing conditions and government priorities, is rare. It affects people, content, controls, processes and technology, and ultimately the underlying business model(s) and relationship with a government's constituency. It is not just about deciding to re-engineer and automate unwieldy processes. Nor is it just about building or buying software, cloud services or hardware through complicated negotiations for licenses. Doing 'digital' upsets the norms and the status quo. It may impact areas that you don't want to touch; create conflict between internal parts of the organization or competition between agencies and departments; force a shift in channels and create new sources of information and citizen/client demand. On top of this, people who have avoided IT, or "don't get IT" become speed bumps to development, capability building and results. This book describes the challenges that confronted Korea and the measures taken—both the successful and the less successful.

The book also captures the experiences and lessons of a succession of governments led by forward thinkers and strong leaders, as well as of private sector partners and committed bureaucrats, who saw (or accepted) technology as both a potential stimulus for private sector development and a lever to create a different type of government relationship with its people. These high-level leaders possessed a wise combination of soft skills, including determination and a willingness to change structures and incentives, and to mediate/negotiate between the silos. Above all, they had an unwavering confidence in their conviction about a technology-enabled future.

Digital disruption has been experienced by a number of industry segments in the last decade. Government is expected to be in the next cohort of industries to be disrupted. Decision-makers in government with last century mindsets about ICT, could benefit tremendously from contemplating the way the Koreans have successfully embraced the digital, as described in the World Bank's World Development Report 2016, and achieved impressive outcomes.

Continuing to move forward despite failures along the way has been an important part of Korea's approach to e-government. This raises the ultimate challenge of going digital in areas/sectors that have relied on 'analog' or manual methods. Certainly, learning from others' successes and mistakes can accelerate the choice of options and actions. And perhaps the desired leapfrogging available to developing nations includes not just the technology, but also the cultural relevance of leadership styles, the upskilling of public sector employees, and dynamic relationships with academic, private sector and civil society organizations, as well as the role of the citizen as partner, not just customer.

Going digital is not an option, nor is it complementary to non-digital government. Government and governance without digital has no future. Many countries and even the 'analog' decision-makers can gain a great deal from reading of the actions and experiences contained in this book. As well as the steps taken, it is important to understand the value of the partnerships created and the technologies applied, and to contemplate the methods and mindset of the leaders whose vision held strong whilst thinking, designing, testing and responding to how government and governance could and should work in the digital age.

Korea's success story is not finished, because government needs to be continually moving forward in terms of technology and becoming more citizen-centric. However, its achievements to date provide valuable guidance for countries that are either already on the digital government path or planning to get onto it.

Jane Treadwell
Practice Manager, Governance Systems, World Bank
Former CIO, Centrelink, Government of Australia

Acknowledgments

This publication was prepared by a joint team from the Korea Development Institute's School of Public Policy and Management, and the Governance Global Practice of the World Bank. The work was led by Changyong Choi (Associate Professor, KDI School), Soonhee Kim (Professor, KDI School), Robert P. Beschel (Lead Public Sector Specialist, World Bank) and Tina George Karippacheril (Senior Public Sector Specialist, World Bank), who edited the publication, in addition to contributing as chapter authors, alongside Jeongwon Yoon (Executive Director, National Information Society Agency, Korea), Jungwoo Lee (Professor, Graduate School of Information, Yonsei University, Korea) and Jooho Lee (Associate Professor, School of Public Administration, University of Nebraska at Omaha, USA).

The team would like to express profound gratitude and deep regard to President Joon-Kyung Kim and Dean Hong Tack Chun of the KDI School, for their support and encouragement to carry out our research on Korea's experience and accomplishments in digital government. We are also grateful to Taejong Kim (Managing Director of Development Research and Learning Network/ Professor, KDI School) for providing insightful and valuable feedback throughout the duration of this research project. Thanks must also go to the KDI School's Development Research Team, comprising Min Young Seo (Development Research Team Head), Youngjoo Jung (Senior Research Associate) and Myung Eun Lee (Senior Research Associate) who offered unflagging administrative support throughout the project. A word of appreciation goes to Eunkyoung Choi (Ph.D. Candidate, Korea University) for her able assistance in collecting data for Chapter 2.

The team would like to thank Jane Treadwell (Practice Manager, Governance Systems), Hassan Cisse (Director, Governance and Inclusive Institutions), and Jim Brumby (Director, Public Sector Performance), Governance Global Practice, World Bank for their leadership, guidance, and steadfast support throughout this project. Thanks to Fekerte Getachew, Laryssa Chiu and Aimee Yuson for outstanding administrative support, and to Graham Colin-Jones for exceptional editorial assistance. A special note of thanks to Grace Porter Morgan (former World Bank colleague, New Delhi Office) for insights and wise counsel throughout the project, and for contributing significantly to the final chapter on lessons learned.

The team gratefully acknowledges the peer reviewers of this publication for detailed and insightful comments at every stage of this project, all the way from the initial concept review to interim, quality, and final output review: Richard Heeks (Professor, Manchester University, IDP), Hee Joon Song (Professor, Ewha Women's University, Chairman of the Prime Minister's Government 3.0 Committee), Cheong-Sik Chung (Professor, Kyungsung University) and Jae Jeung Rho (Professor, KAIST), Tim Kelly (Lead ICT Specialist and WDR 2016 chapter author, World Bank), Samia Melhem (Lead ICT Specialist, World Bank), Robert Taliercio (Practice Manager, East Asia Pacific, Governance Global Practice, World Bank), Cem Dener (Lead Governance Specialist, Global Lead for Integrated Digital Solutions, World Bank), and Zahid Hasnain (Senior Public Sector Specialist and WDR 2016 chapter author, World Bank).

About the Contributors

Robert P. Beschel, Jr., is currently the Global Lead for the World Bank's Center of Government Practice. He has written extensively on policy coordination and public sector reform and worked on Center of Government issues in a diverse number of countries. In 2010, he was recruited by the Office of Tony Blair and the Government of Kuwait to serve as Deputy Director for Policy (and subsequently as Director for Policy) in the newly created Technical and Advisory Office of the Prime Minister. Beschel has managed the Governance Systems Unit and headed the Governance and Public Accountability Cluster in the Bank's Public Sector Anchor. He headed the Secretariat for the Governance and Anticorruption Council—the body that oversees the World Bank's practice on issues of governance, integrity, and anticorruption. He oversaw the World Bank's work on governance and public sector management in the Middle East and North Africa region from 2004 to 2010 and helped to lead the World Bank's governance work in South Asia from 1999 to 2004. He served as the principal author for the Asian Development Bank's Anticorruption Strategy in 1998.

Dr. Changyong Choi is an associate professor at the Korea Development Institute School of Public Policy and Management, South Korea. He also serves as Director of Policy Consultation and Evaluation at the Center for International Development of the Korea Development Institute. He is in charge of Knowledge Sharing Program and various international development projects for countries in Asia, the Commonwealth of Independent States, and Europe. His research interests are governance reform, digital government, private sector and market development, and democratization in developing and former communist countries. His current work explores public-private partnership and the effectiveness of international development cooperation programs. He earned a PhD from the Maxwell School of Syracuse University, a master of public policy and a master of arts in education from the University of Michigan, and a bachelor's degree from Korea University.

Tina George Karippacheril is a senior public sector specialist on digital governance with the World Bank. She recently joined the Social Protection and Labor Global Practice where she is co-leading the SPL Delivery Systems Global Solutions Group. She has over 16 years of experience, working with middle- and low-income countries on modernization programs, cross-agency collaboration, institutional change management, process redesign, innovation, bringing public services closer to citizens, citizen service centers, digital self-service, mobile government and design thinking. From 2011 to 2013, she was based at the World Bank's office in Jakarta, working with the government of Indonesia on public management and technology reforms with the Statistics Agency and the President's Delivery Unit on Open Government and Global Partnership for Social Accountability. She holds a Ph.D. in Technology, Policy and Management from the Delft University of Technology in Netherlands.

Soonhee Kim is a professor of public administration at the Korea Development Institute School of Public Policy and Management. Professor Kim's areas of expertise include public management, human resources management, e-government, and leadership development. She is co-editor of *Public Administration in the Context of Global Governance* (Edward Elgar, 2014), *Public Sector Human Resource Management* (Sage, 2012), and the *Future of Public Administration Around the World: The Minnowbrook Perspective* (Georgetown University Press, 2010). She serves as an editor of international features in the Public Administration Review and as co-chair of the International Institute of Administrative Sciences Study Group on Trust and Public Attitudes. Kim received a Ph.D. in public administration from the Rockefeller College of Public Affairs and Policy at the University at Albany, State University of New York in 1998.

Jooho Lee is an associate professor at the School of Public Administration and an associate director of Global Digital Governance Lab at the University of Nebraska, Omaha. He has been doing research on the antecedents and consequences of information technology adoption by government and citizens, interorganizational/interpersonal networks, citizen participation programs, transparency, and trust in government. His research has appeared in public administration and electronic government journals such as *Public Administration Review, American Review of Public Administration, Administration and Society*, and *Government Information Quarterly*. He earned a PhD in public administration from the Maxwell School of Citizenship and Public Affairs at Syracuse University.

Jungwoo Lee is the director of the Center for Work Science and a professor of information systems and technologies at Graduate School of Information, Yonsei University, Seoul, Korea. His current research focuses on the changing nature of work by information and communication technologies. In the early days of digitalization of government, he has published a developmental model of digital government, providing a theoretical basis for numerous international indices for e-government development. Aside from academic responsibilities, he was a columnist for the Digital Times and the Segyeilbo. In 2013, he ran a news program at the M-Money Broadcasting Station specialized in economic analysis. He holds a PhD and MS in computer information systems from Georgia State University, an MBA from Sogang University, and a B.A. in English language and literature from Yonsei University. Currently, he is involved in International Federation of Information Processing Working Group 9.1 ICT and Work.

Jeongwon Yoon has been working for more than 21 years as the executive director of the National Information Society Agency, Korea. He is responsible for the Information Technology and Policy Assistance Program, assisting more than 40 developing countries. He expanded the program by making partnerships with various international organizations. He also founded Global E-Gov Academy for international capacity building. He successfully launched information and communication technology cooperation centers in nine countries including Mexico and Vietnam. Before this, he was responsible for reviewing the Korean National Finance System, planning the National Backup Center and Digital Certification Authority. He also served as the telecommunication sector coordinator (1998–2000) of the International Y2K Cooperation Center, Auspice of the United Nations. He has bachelor and master degrees in computer engineering from California State University. He has a PhD in information management from Seoul University of Information and Venture.

Abbreviations

ACC	Administrative Computerization Committee
ADSS	Architectural Decision Support System
AIS	administration information system
API	application platform interface
BAI	Board of Audit and Inspection
BcN	Broadband Convergence Network
BPR	business process reengineering
BPS	business process system
BRM	business reference model
CIO	chief information officer
CIOC	Chief Information Office Council
COTI	Central Officials Training Institute
CRM	customer relationship management
DRM	data reference model
EA	enterprise architecture
EDI	electronic data interchange
EDMS	Electronic Document Management System
ETRI	Electronics and Telecommunications Research Institute
FMIS	financial management information system
GDP	gross domestic product
GEA	government enterprise architecture
GEAF	Government Enterprise Architecture Framework
GEAP	government EA portal
GIDC	Government Integrated Data Center
GPKI	government public key infrastructure
HD	high definition
HR	human resource
HRD	human resource development
HTS	home tax service

ICT   information and communication technology
IDI   ICT development index
IETF   Internet Engineering Task Force
IMF   International Monetary Fund
INSC   Information Network Supervisory Commission
IoT   Internet of Things
IPC   Informatization Promotion Committee
IPF   Informatization Promotion Fund
ISP   information systems planning
ITA   information technology architecture
ITAMS   information technology architecture management system
ITRC   Information Technology Research Center
KDI   Korea Development Institute
KGEA   Korean Government Enterprise Architecture
KICS   Korean Information System of Criminal Justice Services
KII   Korea Information Infrastructure
KIPC   Korean Telecommunication Promotion Corporation
KLID   Korea Local Information Research and Development
KMS   knowledge management system
KONEPS   Korea Online e-Procurement System
KT   Korea Telecom
MDR   meta data registry
MHWS   Ministry for Health and Welfare Services
MIC   Ministry of Information and Communication
MIS   management information system
MOCE   Ministry of Culture and Education
MOCI   Ministry of Commerce and Industry
MOGA   Ministry of Government Administration
MOGAHA   Ministry of Government Administration and Home Affairs
MOGLEG   Ministry of Government Legislation
MOI   Ministry of Interior
MOPAS   Ministry of Public Administration and Security
MOPT   Ministry of Post and Telecommunications
MPB   Ministry of Planning and Budget
MST   Ministry of Science and Technology
NAFIS   National Finance Information System
NAK   National Archives of Korea
NBIS   National Basic Information System
NCA   National Computerization Agency

NDMS	National Disaster Management System
NEIS	National Education Information System
NIA	National Information Society Agency
NIFP	National Informatization Framework Promotion
NIPMP	National Information Promotion Master Plan
NTS	National Tax Service
OASIS	Organization for Advancement of Structured Information Standards
OECD	Organization for Economic Co-operation and Development
PaaS	platform-as-a-service
PC	personal computer
PKI	public key infrastructure
PPSS	Personnel Policy Support System
R&D	research and development
RCMS	real time cash management system
RFID	radio frequency identification device
RFP	request for proposal
SCeG	Special Committee for e-Government
SIIS	Social Insurance Information System
SLA	service-level agreement
SMG	Seoul Metropolitan Government
SNS	social networking service
SRM	service reference model
SSIS	Social Security Information System
TIS	Tax Information System
TRM	technology reference model
USN	Ubiquitous Sensor Network

Digital Government in Developing Countries: Reflections on the Korean Experience

Robert P. Beschel Jr., Soonhee Kim, and Changyong Choi

Digital Governance and Development Opportunities

Experts and citizens alike agree that the application of information technology (IT) to the challenges of public administration and effective service delivery has been one of the most powerful and transformative governance trends throughout the developing world. Such "e-governance" or "digital governance" applications began with the computerization of internal government management systems for finance, payroll, and other core government functions.[1] They then spread to information sharing with citizens through web pages and other means of basic outreach and communication, many of which initially flowed in one direction from government to the broader public. By the late 1990s, IT was used to streamline and re-engineer business processes and create "one-stop shops" to facilitate improved service delivery. Websites became more capable and adaptable, serving as two-way channels through which government business could be transacted. Around this time, international organizations, such as the Organization for Economic Co-operation and Development (OECD), European Union, and United Nations, began tracking the use of such technologies, and political leaders from around the world began placing IT initiatives on their national agendas as an important priority, investing considerable resources in developing and implementing e-governance strategies. More recently, the rapid expansion of mobile technologies and cell phone use, which has surpassed 70 percent of the population in countries such as India, is opening up a host of new opportunities to access services and monitor government performance.

In bureaucratic terms, this revolution has unfolded at breathtaking speed. Figure 1.1 captures the change along several dimensions that involve the use of

---

In this book, the terms *e-Government*, *e-Governance*, *digital government*, and *digital governance* are used interchangeably. See the Endnotes section.

Figure 1.1 Trends in Development of PFM System (198 Economies)

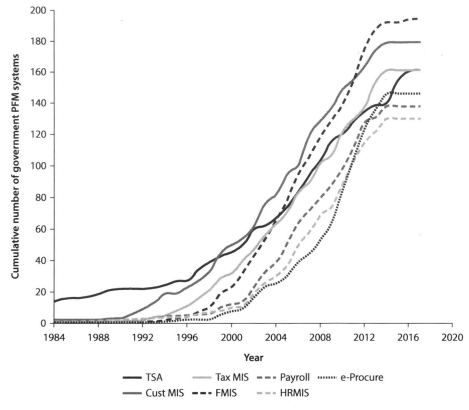

Source: World Bank 2015.

computerized information systems for basic processes within government, with particular emphasis on those involving public financial management (World Bank 2015). Within two decades, such systems (including financial management, customs, taxation, e-procurement, payroll and human resource management) have expanded from a small handful of countries to more than 120. In contrast, the first ombudsman institution appeared in Sweden in 1810, but another 120 years elapsed before the next office was created; as of 1970 fewer than a dozen countries had established such institutions.

Various dynamics have driven this rapid growth. The relentless and well-documented qualitative expansion of computing technology has reduced the relative cost of such systems while greatly expanding their capability. Governments quickly discovered that such technologies could be useful in solving a range of problems, from document processing to coordinating complex workflows that cross organizational boundaries. Under a well-functioning e-governance regime, information and records on the inner workings of governments and policies can be made readily available, thereby promoting transparency and accountability (Brown and Garson 2013; Lee 2010; Song and Cho 2007). In addition, e-governance creates an innovative environment that enhances the efficient delivery and effectiveness of public goods and services (Chen and Dimitrova 2008; Millard 2008). For instance, the application of IT to the government procurement process can streamline its administration and reduce associated costs, as well as creating a more transparent and fairer payment system.

E-governance also improves public disclosure of information on expenditures and policymaking processes, thereby increasing the government's credibility and reducing the possibility of corruption. From a business perspective, companies can invest their resources in more productive ways, since they can spend less time and energy visiting government offices to search for information. In addition, e-governance tends to promote fairness by making information equally accessible to everyone, provided that the problem of the digital divide is addressed. E-governance embodies equity by demonstrating a belief that all members of the public, who are the beneficiaries of government services, should be able to receive administrative information without discrimination and have equal opportunities to become involved in the policymaking process. And unlike the ombudsman office, which is rooted within a particular Western administrative culture and tradition, IT appears more instrumental and less embedded within a given political and social milieu. As such, its application is perceived to be more "value neutral" and has been embraced by countries grounded in both authoritarian and democratic traditions. Whether it actually is value-neutral has been the subject of a broad and ongoing debate.[2]

Not all IT applications have been equally transformative. Existing research has demonstrated that such solutions are typically most effective when applied to tasks and transactions that are routine, predictable, and easily monitored. Complex tasks or those involving a large amount of discretion typically do not see comparable benefits. Information dissemination and the use of online services tend to be skewed toward the young, educated, urban, and financially better-off. Social media, though a powerful tool for drawing attention to blatant and egregious government failures such as corruption or incidents of police abuse, is less effective in identifying ongoing dysfunction in areas characterized by complex causal chains and chronically weak performance.

Even more sobering, many automation efforts within the public sector fail. One analysis of IT projects within developing countries suggested that about 30 percent of them are total failures; 50–60 percent are partial failures, with significant budget and time overruns; and fewer than 20 percent achieve their objectives in terms of time, cost, and functionality (World Bank 2016). Around 26 percent of World Bank-supported IT projects were rated "unsuccessful," as opposed to an average of 18 percent for all Bank projects. Even in OECD countries, government failure rates are significant. One survey of IT projects in the United States reported success rates of 59 percent in the retail sector, 27 percent in manufacturing, and only 18 percent in government (The Standish Group International 2001). In a particularly telling example, Washington, D.C. invested well over $30 million in a failed human resources computerization effort that was ultimately never rolled out.

The Case of Digital Governance Development in the Republic of Korea

How does one reap the benefits of e-governance while minimizing the failures? It is here where the Republic of Korea's experience is particularly impressive. Korea's achievements in field of e-governance have been widely recognized by the international community for the past decade. In 2004, Korea was ranked first in the ITU Digital Opportunity Index. (Ahn 2008; United Nations 2010). In 2005, it was awarded the APEC World Advanced Award for its e-governance system. In 2006, Korea's Online Tax System was recognized by the OECD as one of the best practices in e-government. In 2007, Korea received the U.N. Public Service Award and the e-Asia Award by Asia Pacific Council for Trade Facilitation and Electronic Business (AFACT). It also received the E-Challenge Award in 2008 and has consistently won U.N. Public Service Awards since 2011. Since 2010, Korea has been ranked as the top country in the U.N. e-government survey—a composite index that combines three important dimensions of e-governance: the provision of on-line services, telecommunication connectivity and human capacity (see the breakdown for 2014 in table 1.1).

By any metric, Korea's journey from a devastated and war-torn country in the 1950s, to a developing country in the 1960s, to an advanced information society in the 21st century has been remarkable. As shown in box 1.1, Korea's e-governance evolution can be divided into several broad stages. During the initial "foundation phase" (1980s–1995), the groundwork for e-governance was laid through the digitization of national key databases and by building a network for each government agency. Next came the "full promotion stage," from 1996 to 2002, during which high-speed broadband networks were established across the country and the 11 high-priority IT projects were completed. The third stage of "diffusion and advance" (2003–2007) saw the establishment of government-for-citizens (G4C) applications and the implementation of systems to share administrative information. The "integration stage," from 2008 to 2012, saw the launching of an integrated e-government platform. Finally, the fifth stage of "maturity and co-producing" (2013–2017) is committed to information and communication technology (ICT) innovation for service integration at all levels of government and investment in ICT-enabled growth through working with the private sector and engaging citizens.

Table 1.1 UN e-Government Ranking for Korea

Overall ranking		1
Online Service Index		.98
Telecom Infrastructure Index		.94
Human Capital Index		.93
Extent of service delivery stages (percentage)	Stage 1: emerging information services	100
	Stage 2: enhanced information services	82
	Stage 3: transactional services	77
	Stage 4: connected services	88

Source: United Nations e-Government Survey 2014.

Box 1.1 Korea's e-Governance Experience: A Phased Evolution

1st Stage (1980–1995, Foundation): National Basic Information Systems (NBIS), administrative networks, digitization of national key databases including citizen registration and vehicle registration

2nd Stage (1996–2002, Full promotion): Establishment of nationwide broadband network; completed 11 major tasks for e-government services

3rd Stage (2003–2007, Diffusion and advance): Development of 31 key e-government projects including home tax service, e-procurement, Public Service 24 (Government-for-Citizens, G4C), and administrative information sharing system, etc.

4th Stage (2008–2012, Integration): Integration and joint management of information systems of government agencies; integration and linking of e-government services using cloud computing and hyper connected networks

5th Stage (2013–2017, Maturity and co-producing): E-government 3.0; ICT innovation for service integration; investment in Internet of Things (IoT), Cloud Computing, Big Data for creative economy; ICT-enabled growth and jobs

Korea's e-governance system was introduced for two main reasons. Initially, the focus was on achieving efficiency (and eventually, as the system evolved, greater transparency) in government. Second, the introduction of e-governance was part of a broader national strategy to use technology to shift Korea's economic paradigm from an industrial-based growth economy to a knowledge-based economy and information society.

The system itself can be broadly categorized by the different constituencies served, which include G2G (government-to-government), G2B (government-to-business), G4C (government-for-citizens), and G2C (government-to-citizens). As the following brief discussion illustrates, Korea's progress along each dimension has been impressive.

As in many countries, Korea's early efforts were primarily concerned with achieving greater efficiencies within government (G2G). These included the computerization of basic systems and processes, such as those related to financial and human resource management, and the creation of major databases. They also involved efforts to develop appropriate infrastructure, such as the major push toward high-speed broadband network development. Subsequent efforts have focused more rigorously on ensuring the interoperability of various systems. Since 2011, for example, an integrated e-governance platform has been used by all central government departments and local governments. Under this system, all government-administered work processes such as planning, scheduling, performance management, and decision making are standardized and systematized. In addition, all government decisions are documented and archived, resulting in greater transparency and accountability (Chung 2012; Kwon 2011; Lee 2012).

G2G systems such as the For-Citizen (WiMin) System and the Nationwide Business Process System (On-nara BPS) enable different public organizations to government approached these efforts? Is technology moving in ways that will make Korea's accomplishments easier or more difficult for other countries to achieve in the future?

The remaining five chapters probe various aspects of the Korean experience of e-governance, from initial efforts to build human and institutional capacity to long-term impacts and transferable lessons. Chapter 2, composed by Soonhee Kim and Changyong Choi, addresses the institutional and managerial dimensions of digital governance development within the Korean government. It tackles such important issues as the broader political environment and the nature of leadership from the president and senior ministers; how Korea financed its major IT investments; and how it developed the technical and managerial capacity within the public sector to support its aspirations in this area—including links with e-literacy efforts among the broader public. It also discusses the legal and financial environment within which the e-governance effort unfolded and various institutional mechanisms for coordinating this work within government. Key considerations highlighted in this chapter include the importance of sustained presidential leadership over time; the development of strong managerial capacity for e-governance efforts within the public sector, both at the national level and in project management; the role of collaborative relationships between the public and private sectors; and the benefits of a major push toward developing ICT interest and capacity among the broader public.

Chapter 3, written by Jeongwon Yoon, focuses on building the technical infrastructure for e-governance. It reviews the intimate linkages between Korea's effort to improve government efficiency and its push to use ICT as the basis for gaining national comparative advantage. The chapter highlights many of Korea's major investments in infrastructure, from the early TDX electronic switching device through mid-sized computers and broadband infrastructure to the evolving m-government applications in today's smartphones. It reflects on key facets of these undertakings, such as the virtues of having a well-established IT infrastructure that resolves conflicts among stakeholders; the heavy use of open-source software and strong focus on interoperability, including the creation of a single integrated government data center; and the efficient implementation of short-term projects through specialized public institutions, such as the National Information Society Agency.

Chapter 3 includes a particularly interesting and refreshingly candid discussion of Korean e-governance failures and mistakes. One such error was an inability to identify and capitalize on emerging global market trends due to concentrating too heavily upon domestic technological development and standards. Korea also suffered from redundancies and instances of over-investment in e-governance infrastructure, as well as the hasty inclusion of premature technologies in pilot projects that ultimately did not materialize. A final and particularly important critical finding is the lack of methodical and systematic follow-up evaluation of many IT investments, which could have prevented downstream mistakes and restrained the mainstreaming and scaling up of underperforming projects.

Chapter 4, drafted by Jungwoo Lee, analyzes the evolutionary phases of digital government development in Korea from 2001 to 2012, as the country put in place integrated IT systems and services for digital government. Using the metaphor "from islands to continents," Lee provides an overview of efforts to integrate the 11 key e-governance initiatives and 31 additional priorities under the broader umbrella of the Korean Government Enterprise Architecture (KGEA). The KGEA seeks to assimilate cross-government services into an integrated platform for citizens, businesses, and government agencies; its activity remains ongoing. As of October 2012, 15,000 government systems belonging to 1,400 public institutions had been integrated into the KGEA. Lee also addresses more recent efforts by the Korean government to promote interoperability and "joined-up" government systems. His analysis demonstrates the strong tendency of information systems in Korean digital government to be connected or integrated with each other. He concludes that this synergistic orientation of digital government has affected expectations regarding the continual reengineering of government processes in a way that could not have been imagined before digital government.

Chapter 5, by Jooho Lee, ambitiously incorporates assessments of Korea's e-governance experience written in both English and Korean. This chapter's wealth of information on various impact assessments deserves careful attention from all who are interested in IT's efficacy in facilitating improved performance, accountability, and service delivery. Many of the findings cited in this chapter are consistent with broader experience of e-governance in other countries; for example, routinized tasks can be streamlined, red tape and waiting times reduced, and staff productivity increased. In some agencies, ICT has increased the span of control for middle managers. ICT has made more information available for policy making, and in some instances, end-user satisfaction has improved. Trust in government also appears to be positively impacted. However, other alleged benefits—such as fostering greater social inclusion and cohesion—remain to be seen.

Finally, chapter 6, written by Tina George Karippacheril and others, refocuses the discussion upon the two major objectives driving this endeavor: to succinctly summarize the major lessons from Korea's experience, and to probe their relevance and applicability for other developing countries. The authors distill seven key lessons and apply them to two groups of countries: developing (or low income) countries in need of urgent support to initiate e-Government programs (Group A), and more advanced economies (including middle income countries) that are moving from fragmented information systems to connected platforms (Group B). For Group A countries, the key lessons are: ensuring sustained, high-level leadership and support for digital governance as a national priority; building hybrid technical/functional skills within the public sector; improving interagency collaboration; and sequencing the development of core infrastructure components for a whole-of-government approach that will enhance the provision of services to citizens. For Group B countries, the additional lessons are: empowering local governments to develop a more citizen-centered and service-oriented government; integrating systems to ensure seamless coordination among peer agencies and subnational governments; and establishing public–private partnerships to advance national priorities and achieve better performance outcomes. For each lesson, the authors note the critical success factor and suggest policy implications that are of practical value to the relevant group of countries.

Recognizing that it is equally important to learn from mistakes as well as successes, the authors also examine the measures that did not work and the remedial actions that Korea took to overcome problems that arose. Since technology continues to advance at a rapid pace, the chapter considers opportunities for leapfrogging over less efficient technologies so as to accelerate sustainable development. The chapter concludes with specific implications from Korea's experience that can provide valuable guidance for all countries seeking to develop or advance their Digital Governance program.

On behalf of both the Korean Development Institute and the World Bank, it is our collective hope that the insights captured within this volume will stimulate a creative and constructive debate, both within Korea and in many other countries, as to what we can learn from Korea's experience and how we can best apply those lessons in a variety of local contexts. We believe that Korea's experience, when appropriately considered and distilled—and when carefully aligned with local circumstances and capacities—offers much to others interested in the successful application of IT to the challenges of improving transparency, accountability, and service delivery throughout the public sector. In many ways, Korea is helping to redefine what is possible in the area of digital governance. Its experience, both good and bad, is inspiring and highly instructive.

Notes

1. Some scholars argue there are analytic distinctions between "e-Governance" and "Digital Governance". The former emerged earlier in the late 1990s, where the "e" stood for electronic communication. Digital Governance gained currency a decade later and refers to a broader and more holistic understanding of the way in which countries apply technology to improve governance outcomes. However, in this volume, the terms are used interchangeably.

2. Some scholars have argued that e-governance ultimately fosters public participation in policy making. See Kim and Lee 2012; Macintosh, Gordon, and Renton 2009.

Bibliography

Ahn, M. S. 2008. *The Theory of E-Government in Korea*. Seoul: Bak-Yeong Press.

Brown, M., and G. D. Garson. 2013. *Public Information Technology and e-Governance: Managing the Virtual State*. Hershey, P.A: IGI Global.

Chen, Y.-C., and D. V. Dimitrova. 2006. "Electronic Government and Online Engagement: Citizen Interaction with Government via Web Portals International." *Journal of Electronic Government Research* 2 (1): 54–76.

Chung, C. S. 2012. *The Theory of Electronic Government*. 3rd ed. Seoul: Seoul Economy and Business Management.

Garson, G. D. 2006. *Public Information Technology and E-Government: Managing the Virtual State*. Sudbury, MA: Jones and Bartlett Publishers.

Heeks, R. 2001. *Reinventing Government in the Information Age: International Practice in IT-Enabled Public Sector Reform*. London: Routledge.

Kim, D. J. 2010. *Kim Dae-Jung's Autobiography*. Seoul: Sam-In Press.

Kim, S., and J. Lee. 2012. "E-Participation, Transparency, and Trust in Local Government." *Public Administration Review* 72 (6): 819–28.

Kim, S. J. 2011. "The Success Factors and Future Tasks of the Korean e-Government." *Korean Balanced Development Studies* 2 (3): 55.

Kwon, G. H. 2011. *The Theory of E-Government*. Seoul: Pakyoungsa. (in p. 39)

Lee, J. 2010. "10 Year Retrospect on Stage Models of e-Government: A Qualitative Meta Synthesis." *Government Information Quarterly* 27 (3): 220–30.

Lee, Y. B. 2012. "2011 Modularization of Korea's Development Experience: The Introduction of E-Government in Korea." Korea Development Institute (KDI) School of Public Policy and Management research paper. Seoul: Ministry of Strategy and Finance (MOSF).

Macintosh, A., T. F. Gordon, and A. Renton. 2009. "Providing Argument Support for E-Participation." *Journal of Information Technology & Politics* 6 (1): 43–59.

Millard, J. 2008. "E-Government for an Inclusive Society: How Different Citizen Groups use E-Government Services in Europe." In *E-Government Research: Policy and Management*, edited by D. Norris. Hershey, PA: IGI Publishing.

MOSPA (Ministry of Security and Public Administration). 2014. *Committee of Government 3.0*. Government 3.0 Development Plan Report, Committee of Government 3.0. Seoul: MOSPA.

O' Looney, J. 2002. *Wiring Governments: Challenges and Possibilities for Public Managers*. Westport, CT: Greenwood Publishing Group.

Otenyo, E. E., and N. S. Lind. 2011. *E-Government*. Amherst, NY: Teneo Press.

SCEG (Special Committee for E-Government). 2003. *2003 White Paper of E-Government*. Seoul: SCEG.

Song, H. J., and T. Cho. 2007. "Electronic Government of Korea: Performance and Tasks." *Informatization Policy* 14 (4): 20–37.

The Standish Group International Inc. 2001. "Extreme Chaos, 2001." http://www.cin .ufpe.br/~gmp/docs/papers/extreme_chaos2001.pdf.

United Nations. 2010. *E-Government Survey 2010*. New York: United Nations.

———. 2014. *E-Government Survey 2014*. New York: United Nations.

World Bank. 2015. *Doing Business 2016: Measuring Regulatory Quality and Efficiency*. Washington, DC: World Bank.

———. 2016. *World Development Report 2016: Digital Dividends*. p. 165. Washington, DC: World Bank.

Yoon, Y. M. 2003. "Leadership and Coordination of the E-Government: Lessons from the People's Government." *Quarterly Thought* 57 (Summer): 55–74.

CHAPTER 2

Institutional and Managerial Dimensions of Digital Government Development in the Republic of Korea

Soonhee Kim and Changyong Choi

Introduction

The Republic of Korea's journey from a developing country in the 1960s to an advanced information society in the 21st century has been a remarkable one. The e-government system was initiated in Korea for two reasons: (1) to reform the civil service and improve transparency in government; and (2) to shift the economic paradigm from an industrial-based growth economy to a knowledge-based economy and an information-based society, through technology. The focus was not just on adapting to a new technology; rather, it was on bringing greater efficiency to government services and enhancing the delivery of public services to citizens. Korea's achievements in this regard have been widely recognized by the international community through various awards from prestigious bodies between 2005 and 2011.

This chapter explores two key questions:

1. What are the institutional and managerial factors that have facilitated successful e-government development in Korea?
2. What are the key components and practices of e-government leadership, finance and capacity building that have contributed to effective Korean e-government development?

The journey of e-government development in Korea suggests that such a transformation requires a high degree of political willingness and commitment to support the national agenda of e-government under a long-term strategy of development (Fountain 2004; Heeks 2001; O'Looney 2002; Song 2004). This shift also demands a strategic approach that enables cross-jurisdictional collaboration and alliance, matrix or virtual organizations, reengineering of business operation processes, integration of public services, human resources development, and constant monitoring of feedback to foster further enhancements. To move toward efficient and effective citizen-centered service delivery, an e-government development strategy also requires civil servants to be committed and motivated to change bureaucratic structures and transaction processes through coordination and collaboration among various agencies.

In examining Korea's e-government development experience, this chapter analyzes the following aspects: (1) the role of formal institutions (i.e., laws and agencies) and e-government policy design; (2) versatile political and managerial leadership in e-government development; and (3) finance and capacity building of human capital. Finally, the chapter addresses the challenges of e-government development in the context of Korea and considers the implications of Korea's achievements and challenges for e-government in developing countries.

Institutions and Digital Government Policy Design: A Brief History

Establishment of Institutions in the Early Stages of IT Policy Design
ICT Initial Stage (1960s and 1970s)

Establishment of the correct formal institutions and creation of the appropriate e-government policy design at an early stage were critical factors in the eventual successful implementation of the e-government program (Chung 2012; Kwon 2011; Lee 2012b).

Before 1978, the development of a sustainable knowledge- and information-based society—informatization—insofar as it related to administrative services, was implemented by individual ministries. The informatization was undertaken to address weak capacity in public institutions, as reflected by the low distribution rate of personal computers (PCs) in public organizations and the significant amount of work conducted on typewriters. Despite the lack of physical infrastructure for the IT industry, the Korean government recognized that invigorating the information and technology industry was necessary for increasing its international competitiveness.

Since there was no computer industry in Korea, the government established a committee for computer development under the Ministry of Science and Technology (MST) and tried to introduce computers into government administration. Individual ministries made attempts to computerize administrative tasks, with the success of implementation largely depending on the circumstances of each agency. Therefore, the effectiveness of the computerization was low.

Noting the lack of progress, President Park Chung-hee ordered that the administrative computerization be led at the national level by the central government. Under the Park administration, the government transferred the Central Department of Computing, which was under the MST, to the Governmental Department of Computing under the Ministry of Government

Administration (MOGA). MOGA established an Administrative Computerization Committee (ACC), together with the ground rules and basic plans to implement computerization in government administration at the national level.

The First Five-Year Basic Plan for Administrative Computerization (1978–82) was the first national policy that integrated individual agencies' computerization projects under MOGA's direction.

Foundation (1980–1995)

In 1982, the Second Five-Year Basic Plan for Administrative Computerization (1983–87) was formulated to integrate the computer network not only among central government agencies, but also at the local government level. Both the plans and the computerization projects in the initial stage of e-government implementation were carried out by MOGA and the ACC.

The Second Five-Year Basic Plan included the National Basic Computing Network Project, which emphasized the development of e-government in various sectors such as administration, finance, education, research, and national defense. The goal of the project was to make Korea an information society at the level of advanced countries by the year 2000. This plan also included the government's desire to realize efficient government, improve convenience for users, and promote the productivity of companies, by expanding the computer network so as to ultimately secure and maintain national competence in information technology. In particular, computerization of administrative tasks would allow frontline users to improve efficiency in carrying out their tasks.

The legal framework for the National Basic Computing Network Project was based on the 1986 Act on Establishment and Utilization of Network (see table 2.1). To accomplish the National Basic Computing Network Project, the Administrative Computing Network Plan was established. This plan, which divided accomplishment of the goals among related government agencies, assigned the function of a computing center to the Department of Computing in MOGA.

Subsequently, the Computing Network Steering Committee was created in 1987 to deliver and coordinate programs and activities related to the computing network. Initially, it was supervised directly by the President, but in 1989 responsibility for informatization was transferred to the Ministry of Post and Telecommunications (MOPT) (Chung 2012). The Minister of MOPT was appointed as the chairperson, and vice ministers and heads of relevant government agencies, such as the Economic Planning Board, Ministry of Finance, MOGA, Korea Bank, and the National Computerization Agency (NCA), were the members of the steering committee.

The NCA was established in 1987 to inspect and provide technical support on e-government projects. Before the NCA was created, the role of inspecting e-government was assigned to the Board of Audit and Inspection (BAI), but the Administrative Computing Network project could not be completed satisfactorily because BAI did not have expertise on e-government projects. This led the

Bringing Government into the 21st Century • http://dx.doi.org/10.1596/978-1-4648-0881-4

Table 2.1 History of Korea's E-Government Implementation

Stage of e-government	Main accomplishments and laws
ICT initial stage (1960s and 1970s)	Introduction of computers to the statistics work of the Economic Planning Board (1967)
	Establishment of the First Five-Year Basic Plan for Administrative Computerization (1978)
1st stage (1980–1995): Foundation	National Basic Information System (NBIS), administrative networks, digitization of national key databases including citizen registration and vehicle registration
	Formulation of the Second Five-Year Basic Plan for Administrative Computerization (1982)
	Plan to distribute multi-functional office equipment (PC) (1986)
	The Computing Network Act (1986)
	Act on Establishment and Utilization of Network (1986)
	The National Basic Computing Network Project and Administrative Computing Network Plan (1987–1991)
2nd stage (1996–2002): Full promotion	Establishment of a nationwide broadband network
	The 1st and 2nd National Informatization Promotion Master Plans (1996–1998, 1999–2000)
	Informatization Promotion Act (1996)
	Digital Signature Act (1999)
	Formulation of the First E-Government Plan and Implementation of 11 e-government projects (2001–02)
	E-Government Act (2001)
3rd stage (2003–2007): Diffusion and advance	Formulation of the Second E-Government Plan and development of 31 key e-government projects including Home Tax Service, e-Procurement, Public Service 24 (Government-for-Citizens, G4C), Administrative Information Sharing system
	Implementation of the 31 e-government projects (2003)
	Preparation of the groundwork for linking and integrating government institutions and departments (2003–07)
4th stage (2008–2012): Integration	Integration of information systems of government agencies; integration and linking of e-government services using cloud computing and hyper connected networks; expanded administrative information sharing; implementation of 12 e-government tasks for openness, sharing, and connection and cooperation
	Establishment of National ICT master plan (2008)
	The Framework Act on National Informatization (2009)
	Act on Shared Utilization of Public Administration Information (2010)
5th stage (2013–2017): Maturity and co-producing	E-government 3.0; ICT innovation for service integrations; investment in Internet of Things (IoT), Cloud Computing, Big Data for creative economy; ICT-enabled growth and jobs
	Act on Promotion of the Provision and Use of Public Data (2013)

Source: Adapted from Lee 2012b; Song 2004; Song and Cho 2007; Special Committee for e-Government 2003; MOSPA 2014.
Note: ICT = information and communication technology.

government to launch the NCA, under the direction of MOPT, to monitor and standardize relevant technologies to carry out the e-government project. In 1994, the MOPT was reorganized into the Ministry of Information and Communication (MIC) and became the central authority in the information and communication technology (ICT) industry.

Institutions for Action: From Full Promotion to Maturity
Full Promotion (1996–2002)

During the full promotion stage of e-government in Korea (1996–2002), the Korean government established a nationwide broadband network and completed 11 major tasks for e-government services (see table 2.1).

Important developments during this stage were:

- *The Informatization Promotion Act.* This came into force in 1996, providing the necessary legislation for implementation of an informatization plan for the public sector;
- *The first National Information Promotion Master Plan (NIPMP) (1996–1998).* Formulated by MIC, the first NIPMP contained detailed plans on how to connect public sector data with other sectors through the existing informatization network, and guidance on how to use such data (Chung 2012).
- *E-Government Vision and Strategy.* In 2000, the Ministry of Government Administration and Home Affairs (MOGAHA, formerly MOGA) formulated the E-Government Vision and Strategy, which laid out the detailed development steps for e-government that were to be completed by 2002. While MOGAHA implemented the E-Government Vision and Strategy, MIC proposed detailed e-government activities by formulating Cyber Korea 21, but the project did not have a long-term vision or a systematic approach due to lack of interagency coordination (Lee 2012b).
- *The E-Government Act (2001).* The most important institutional justification for promoting e-government development was the E-Government Act of 2001, which provided the national vision for e-government development. Article 1 of the Act provided the broad objective of e-government development in the context of governance (i.e., public administration and citizens) by stating that the purpose of e-government is to enhance citizens' quality of life by increasing the productivity, transparency, and democracy of administrative agencies (Soh 2003; Song 2004).
- *The Informatization Promotion Fund.* This was created and used to support companies and enterprises that introduced and developed IT equipment and software that were necessary for e-government.

In the full promotion stage, e-government plans laid out by MOGAHA and the MIC overlapped, as the two organizations competed with each other to gain the initiative and control of the e-government policy (Hwang 2000). This competition led to the establishment of the Informatization Promotion Committee to mediate and coordinate between the two ministries.

Recognizing the challenge of interagency coordination, in 2001 the government established the Special Committee for e-Government (SCeG) under the direct supervision of the Presidential Committee on Governmental Innovation (Soh 2003). The SCeG, which reported directly to the President via the Senior Presidential Secretary for Policy Planning, was autonomous and had discretion in the actual operations of e-government projects. The SCeG chose 3 key priorities: (1) improving service for citizens and businesses (front-end); (2) enhancing administrative productivity (back-end); and (3) establishing infrastructure. Based on these priorities, the First E-Government Plan was devised in 2001, with 11 e-government activities.

Diffusion and Advance (2003–2007)

Upon completion of the First E-Government Plan during the Kim administration, the Roh Moo-hyun (2003–2007) administration drafted and implemented the Second E-Government Plan with 31 e-government activities in four areas (Lee 2012b). During the Roh administration, the E-Government Professional Committee was established under the Presidential Committee on Governmental Innovation and Decentralization (renamed from the Presidential Committee on Governmental Innovation), and operated from 2003 to 2005. However, due to the restricted institutional and legal status of the committee, it did not play a vital role in pushing forward important e-government projects. The E-Government Professional Committee was abolished in 2005, and the SCeG was reorganized under the Presidential Committee on Governmental Innovation and Decentralization. A few months later, the role of the SCeG was transferred to MOGAHA because the government thought that the ministry had stronger enforcing power to implement and manage e-government projects than the SCeG (Lee 2012b).

Integration (2008–2012) and Maturity and Co-Producing (2013–2017)

During the Lee Myung-bak administration (2008–2012), the government abolished the MIC, and its role in informatization was transferred to MOGAHA. In addition, the Presidential Committee on Governmental Innovation and Decentralization was abolished, while the President's Council on Information Strategies was established under the Framework Act on National Informatization of 2009.

From 2009 to 2012, the Korean e-government was focused on establishing an "advanced stage" in which an integrated e-government platform was launched. Furthermore, from 2013 to 2016, the Korean government has been committed to ICT innovation for service integration at all levels of government, while investment in ICT has brought growth and jobs focusing on Internet of Things (IoT), Cloud Computing, and Big Data utilization (Ministry of Security and Public Administration [MOSPA] 2014).

Special attention should be paid to the Korean government's ongoing commitment to digital government. Under 'Government 3.0', the Korean government has emphasized the establishment of 'Service-oriented Government, Capable

Government, and Transparent Government' as the goals of digital governance. These goals are to be achieved through the core values of openness, sharing, connection and cooperation. Notable achievements to date are the reduction of working days for civil petition from 20 days to 7.4 days by utilizing a One-Stop Petition Service System and the construction of big data to provide the public with employment information and social welfare programs (MOSPA 2014). More specifically, through the collaboration of several agencies, there is now an expanded communication channel between the government and citizens, resulting in enhanced public service delivery through mobile phones, and the use of Social Networking Services (SNS) to inform citizens of policies (MOSPA 2014).

Challenges of Developing Local E-Government under a Centralized Regime

The legal foundation and centralized institutions for e-government policy design and implementation have had a positive influence on e-government development in Korea. However, scholars acknowledge that the top-down and centralized e-government development brought some challenges for local governments (Song 2004; Song and Cho 2007). Adopting a centralized approach was beneficial in developing and implementing e-government projects, but it did little for flexibility and a bottom-up approach from local governments in the early stage of e-government development in Korea.

When the local governments initiated and implemented e-government projects, mayors and district officers faced many difficulties because of Korea's top-down organizational culture. Even though the national government granted autonomy to all local governments (followed by the Self-Governance Act of 1988), a hierarchical intergovernmental relationship between central and local government still existed, and districts still needed institutional and legal approvals from upper-level governments. Therefore, to implement an e-government project, mayors and provincial governors had to visit MOGAHA, which was in charge of issuing regulations; the National Assembly and its subsequent committees that were handling issues related to e-government and its budget; and the Ministry of Planning and Budget (MPB) that made sure projects and budgets were implemented properly and according to the annual plan. Districts received annual audits from BAI, and these audits prevented them from using the budget and designing the program with flexibility.

Organizational problems arose not only at the central government level; at the local or district level, mayors and provincial governors also faced challenges with their employees. Mayors wanted to implement e-government to improve efficiency and transparency in their administrations, but employees worried that e-government systems and technology would reduce jobs rather than create them. Employees, worried about their job security, had no channel to address this issue, and it was difficult for mayors to get consensus on delivering e-government projects (Bretschneider et al. 2005). As will be discussed more in the following sections, the national and local governments had to develop human resource development (HRD) programs to increase the usage of information technology, which then would further strengthen the job security of civil servants.

Bringing Government into the 21st Century • http://dx.doi.org/10.1596/978-1-4648-0881-4

Leadership in Digital Government in Korea

A key success factor for e-government development in Korea was the leadership of the SCeG, which took a leading role in directing 11 e-government projects in 2001 and 2002 (Song 2004). Other important success factors were the Korean government's investment in a capacity-building strategy to connect e-government innovation to government reform; strong collaboration; project management capacity; finance; and HRD for ICT (Soh 2003; Yoon 2003). To explain these factors, this section applies leadership theories or models that emphasize the links among organizational leadership, change or reform, and management capacity-building.

First, the integrative leadership model proposed by Ingraham, Sowa, and Moynihan (2004) emphasizes an integrated perspective concerning leadership and management capacity of human resource management, financial management, infrastructure management, and information technology. The model further articulates the feedback process of an integrative leadership, which takes project or innovation management as the form of an ongoing organizational learning process. Second, this study takes transformational leadership theory as a way of understanding the leadership role of committees as change agents that initiate and implement new directions within organizations. Overall, transformational leaders manage change in organizations through a three-act process, including the recognition of the need for change, the creation of a vision, and the institutionalization of change (Tichy and Devanna 1986). The most challenging task of the transformation leadership is to institutionalize change; transformational leaders need to replace old structures with new ones to implement new visions and ideas. Finally, concerning implementation of the new vision and ideas such as e-government innovation, Moore (1995) proposes the application of three elements of corporation strategy for creating and implementing public value that managerial leadership should consider: (1) develop a substantively valuable mission and goals for stakeholders; (2) promote a legitimate and politically sustainable strategy; and (3) build a feasible operational strategy. These perspectives are applied to analyze the e-government transformation in Korea.

Political Leadership

Scholars acknowledge that one of the most important success factors of Korean e-government is the three decades of presidential leadership which supported it. Consistent executive leadership support was necessary for funding, coordination among agencies, and the revision of laws and rules (Song 2004; Song and Cho 2007; Yoon 2003).

The accomplishments listed in table 2.1 provide evidence of over 30 years of commitment to e-government development under different administrations. The First Five-Year Basic Plan of Administrative Computerization was implemented under the second Park Chung-hee administration (1978–82), and the Second Five-Year Basic Plan (1983–86)—the first comprehensive project for e-government in Korea—during the Chun Doo-hwan administration (Lee 2012b; NIA 2005). The Chun administration announced 1983 as the "Year of the Information Industry" and initiated the National Basic Information System (NBIS) project under the leadership of the Presidential Information System Committee. The Chun Doo-hwan administration established several laws that facilitated IT industry development during the 1980s. Although President Chun did not have expertise in IT, he paid attention to the techno-crats who emphasized the benefits of computerization in the near future (Lee 2012b), and he supported the work of the National Basic Computing Network Project and the Administrative Computing Network Plan (Lee 2012b; Song and Cho 2007).

Inspired by the U.S. information superhighway in 1993, the Kim Young-sam administration established the MIC at the end of 1994 and implemented the High-Speed Broadband Network Project through the IT Development Committee, established under the prime minister's office (Song 2004). The high-speed broadband network was essential for expanding e-government innovation throughout the nation. By the end of 2000, the Kim Dae-jung administration had established high-speed Internet broadband in 144 regional areas.

After the Asian Financial Crisis in 1997, the Kim Dae-jung administration (1998–2003) implemented market-oriented reforms, such as deregulation and privatization. Additional reforms focused on the institutionalization of transparent governance and e-government development to promote openness, participation, and integrity. President Kim presented the building of an information-knowledge society as a national agenda, focusing on the institutionalization of transparent governance through proactive e-government development to promote openness, participation, and integrity (Kim 2010; Kim 2011; Yoon 2003). This section focuses on the Kim Dae-jung administration because his administration fully promoted e-government development. Moreover, he made the connection between developing e-government and moving public administration toward more transparency, integrated online services, and less corruption in government (Song 2004; Song and Cho 2007).

Immediately after settling on a government reform strategy for overcoming the Asian Financial Crisis, President Kim Dae-jung established the SCeG in January 2001, with a public–private partnership approach (Song 2004). In 2001 and 2002, the committee successfully completed 11 e-government projects, including the following:

- An e-procurement service system. This began in September 2002, allowing private companies to obtain procurement information through the electronic system and participate in competitive bids.
- A citizen service portal. By November 2002, citizens could access electronically 4,000 types of civil matters and were able to download 393 types of different documents (Kim 2010).
- System for information sharing. Different branches of government were able to access the same information and share information with each other.

President Kim's initial motivation to promote e-government was ascribed to discussions with Alvin Toffler, Bill Gates, and Masayoshi Son about e-government, the information society, and IT education for students and citizens (Kim 2010). His interest and desire to promote e-government was expressed on many occasions, from the inauguration speech to an ordinary cabinet meeting. He encouraged public servants and citizens to enhance IT knowledge and build an information-oriented society as a way of enhancing national competitiveness and sustainability. His vision for the Korean e-government was elaborated in his biography in 2010:

> This society would increase efficiency, demolish corruption, and improve public trust in government if an e-government system is established successfully in our country. Entrepreneurs would run business in a more business friendly environment, and Korea would be one of the most competitive countries in the world. It cannot be too much to emphasize the significance of e-government. Let's make Korea as one of the most advanced countries in the world by establishing an e-government system successfully (Kim 2010, 444).

President Kim also held a strategic meeting on informatization with cabinet agency secretaries and issued a presidential order to create a position of Chief Information Officer (CIO) in government agencies at the national level. He also had regular meetings with his senior secretary for policy planning, who was a member of the e-government committee, to check the e-government committee's project progress. The President reconfirmed his commitment to speedy e-government development, especially the 11 projects under the SCeG (SCeG 2003). Sometimes, Ahn Moon-suk (the Chairman) reported to the President in person (Song 2004).

President Kim's exceptional interest became a driving force, which was necessary for the process of coordinating different opinions among the government agencies and for institutionalizing the value of innovation in the public sector (Garson 2006; Otenyo and Lind 2011; Titah and Barki 2008). The President's strong will made it possible to allocate the budget on time to the 11 projects of the First E-Government Plan and to establish cooperation among the agencies (SCeG 2003). In his autobiography, President Kim addressed the challenges of information sharing coordination among agencies:

> Ministries established databases and the central government modified certain IT-oriented laws and systems, such as the Framework Act on Informatization Promotion, E-Sign Act, and Electronic Government Act. However, it was still challenging to tear down the walls between/among ministries. Ministries refused to share information with other ministries. It was a very closed-door administration system. I ordered the e-government committee to carefully assess these situations and to report any difficulties to me (Kim 2010, 444).

President Kim not only made extensive efforts to build an Internet infrastructure, but he also invested in IT education for students and citizens to establish an information civil society with active utilization of e-government by all citizens.

The Korean government connected high-speed Internet service to all elementary and junior high schools in December 2000, and provided PCs to 330,000 teachers. The President also paid attention to issues of the digital divide. Free computer education was delivered to 500,000 students from lower-income families, as well as free computer training for citizens with low household income, targeting informatization training programs for 10 million citizens by March 2001 (Kim 2010). As a result of these efforts, Kim's administration completed the establishment of the e-government foundation in November 2002. In 2001, Korea ranked 15th in the UN E-Government Readiness Index, and 13th in 2002 (SCeG 2003).

President Roh Moo-hyun expanded the e-government project further in 2003 through support from civil society (Song 2004). Like President Kim Dae-jung, the Roh administration (2003–08) gave its support to MIC to take a leading role in creating a vision for the goals of e-government and in achieving those goals. MIC was also responsible for managing the Informatization Promotion Fund (IPF). President Roh demonstrated his commitment to e-government development by participating in important decision-making processes whenever necessary. For example, he resolved conflicts related to system redundancy when new systems were developed (i.e., the On-nara system, which was developed to support government business, and the existing e-document system, E-nara). Moreover, when there was conflict between MOGAHA and MIC over the National Computing and Information Agency initiative, President Roh himself mediated and coordinated efforts to resolve the situation (Lee 2012b).

In summary, the development of presidential vision, a strong commitment to delivery, and leadership skills for coordination have been the greatest driving forces in the pursuit of the e-government agenda in Korea.

High-Level Leadership for Coordination

One of the most challenging tasks of e-government leadership is coordinating e-government projects among government agencies to ensure interoperability, avoid duplication, ensure coherent action in a range of crucial areas such as security and privacy, and provide the framework and capacity for seamless services (OECD 2003). An effective organizational coordination process for e-government development demands a participatory design, stakeholder analysis, determination of priorities and issues across agencies, and respect for the different levels of technical maturity and identity.

Since the early stages of e-government development, the Korean government understood the importance of a dedicated organization that could take charge of e-government project coordination and make mid- and long-term master plans for e-government development (Lee 2012b). As mentioned earlier, the need for a coordinating organization among ministries was raised often and led to the establishment of the SCeG in 2001 (Soh 2003). This committee, however, was a temporary entity under the Presidential Committee on Governmental Innovation, which was led by the MPB. The budget for the projects that the SCeG engaged in was submitted to the National Assembly without reduction by MPB. This was one of the reasons why the ministries acknowledged the authority of the SCeG and showed their willingness to cooperate with the selected e-government projects (Soh 2003).

The SCeG consisted of 17 members from the public and private sectors: the senior presidential secretary for policy planning, seven vice ministers and two vice minister-level officials from the government, and seven experts from the private sector including the chairman, Ahn Moon-suk, a Korea University professor (Yoon 2003). The combination of the committee members from government agencies, the presidential office, the public sector, and universities brought diverse talents and competencies to lead the management of the national level e-government project. There was a general perception in the Korean government that a committee has weak power when a civilian takes the head position (SCeG 2003). However, scholars found a different power dynamic in this committee when the opinions and interests of related ministries were in sharp conflict (Soh 2003; Yoon 2003). The civilian chairman functioned effectively to ease the tension and narrow the gaps between ministries (Yoon 2003).

Another component of the success of the SCeG was the strategic decision to prioritize the First E-government Plan and implement the 11 e-government projects under the committee's direction. Two committee members from the private sector were allocated to each activity, and several teams were organized and run as necessary (see figure 2.1; Anh 2008; Lee 2012b). Like in the integrative leadership model (Ingraham, Sowa and Moynihan 2004), the strategic and

Figure 2.1 Structure of the e-Government Project during the Full Promotion Stage (1996–2002)

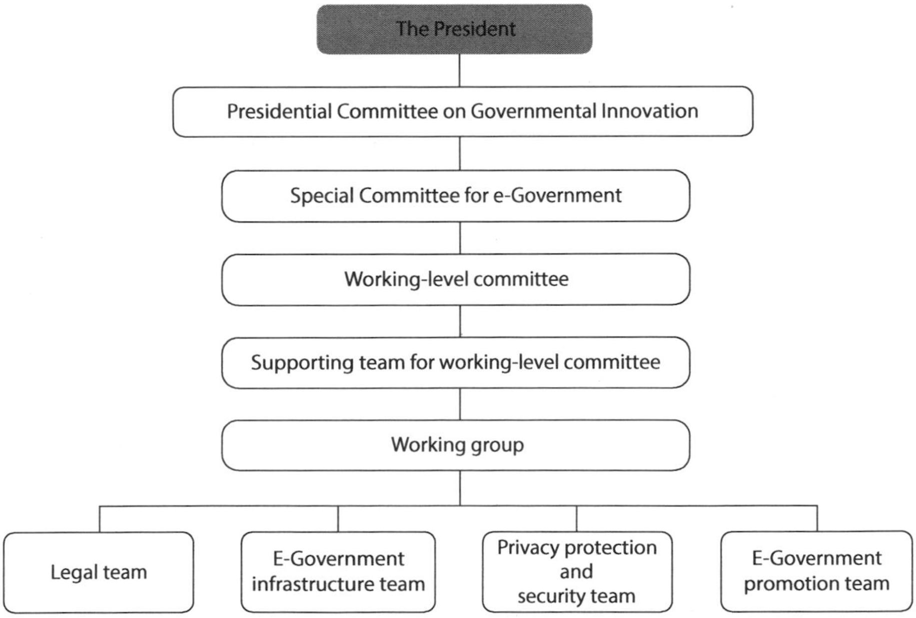

Source: Special Committee for e-Government 2003, 59.

selective decisions on the e-government projects provided an opportunity for committee members to demonstrate their capacity for managing for results.

The SCeG designed its structure and communication channels to coordinate the different opinions among the ministries and to secure technical support. Two co-chiefs of the working-level committee were appointed: the chief of the NCA for technical support and the Presidential Secretary for Policy for coordination among ministries (Ahn 2008; SCeG 2003). The SCeG held a meeting of the working-level committee members every week, where they monitored the proceedings of the project, made policy decisions, and discussed the problems arising in the course of implementing the project (Ahn 2008). The important issues discussed in the meeting were reported directly to President Kim Dae-jung by the Senior Presidential Secretary for Policy Planning (Ahn 2008; Yoon 2003).

In addition, each civilian committee member was responsible for reviewing and coordinating the activities of the e-government projects. Only when a conundrum arose beyond their capacity did the chairman step in. If the problem could not be resolved with intervention of the chairman, it was placed on the table in the plenary meeting that was held every one or two months (SCeG 2003). This approach demonstrated that empowered leadership and ownership of the committee members for project management was resulting in effective coordination.

Another of the committee's operational strategies was to assign responsibility for a certain activity to its implementing ministry (SCeG 2003). Specific duties and responsibilities for selective e-government projects for a government agency were assigned to the vice minister members, who were asked to report directly to the President the activities related to their ministry. When the vice minister reported to the President, the minister was also present to hear the President's comments and instructions (Ahn 2008). The public relations for e-government projects were also handled by each ministry in charge, not by the committee. In sum, by having the ministries take responsibility for their own projects and public relations, the committee could play an effective role in ensuring coordination among agencies. This strategy demonstrates how the empowerment approach to project management fostered successful e-government development in Korea.

During the course of e-government project implementation under the SCeG, three issues were assigned particular importance:

- *Securing financial resources.* The IPF, which was quite flexible, was utilized, but this would not have been possible without cooperation and support from MIC and NIA.
- *Amending laws and regulations.* A taskforce that reported to the SCeG was formed to get cooperation from the National Assembly, which is needed to amend e-government-related laws. At the same time, the committee members made contact with the opposition party's members individually and explained the importance of the project.
- *Standardizing technology.* The SCeG set up a taskforce that included many private sector experts, given that technological advances were very rapid.

Overall, both the leadership of the chairman (including his expertise and impartiality) and the dedication of the committee members contributed to improved performance and coordination among ministries (Song 2004). Chairman Ahn was an expert in administration and informatization, and had been involved in government affairs for more than 30 years, gaining the respect and trust of many of the bureaucrats (SCeG 2003). He also had a passion and drive for the project, attending early morning meetings for more than 2 hours every week (Yoon 2003). His leadership stimulated and motivated public officials to actively participate in the project. The other committee members were also critical to the success of the project. Their dedication is demonstrated by their spending more than one day a week on the project on average and attending 51 working-level meetings and 8 plenary sessions, along with publishing 3 presidential reports (Ahn 2008; SCeG 2003).

Song (2004) states that the expertise and dedication of the committee members inspired the officers of the ministries, causing them to respond to requests from the Committee, stay actively engaged in the project, submit all required data and materials, and express their opinions and ideas (SCeG 2003). Thanks to the committee members' dedication and professionalism, the Committee was evaluated as objective and impartial in terms of addressing conflict or different opinions among the ministries. Effective communication by the Committee's leadership also helped reduce and resolve conflicts that arose during the implementation process (Chen and Dimitrova 2008; Millard 2008).

Accordingly, it can be concluded that the SCeG provided the necessary leadership for coordination and established a strong foundation for the expansion and successful development of e-government in Korea.

Project Management: Operational Leadership and Collaboration

Operational leadership for effective management has been identified as a success factor of the Korean e-government development (Lee 2012b; Song 2004; Suh et al. 1996). To understand the project management strategy for Korean e-government development, it is necessary to revisit the full promotion stage of Korea's e-government development (1996–2002). This stage focused on building the integrated administration network among government agencies and establishing online public service delivery, following the increased diffusion rate of computers and the wide availability of a high-speed information network (Lee 2012b). During this time, operational leadership for project management was applied to the e-government project, including the forecasting analysis of future ICT service demands, stakeholder analysis, feasibility assessments, and a roadmap for prioritizing short-term and long-term projects.

The Korean government adopted a top-down approach to implement the first Administrative Computing Network Plan by prioritizing e-government projects listed by the Computing Network Steering Committee rather than by each agency (1987–91) (Lee 2012b). After enacting the Act on Establishment and Utilization of Network in 1986, the Korean government established the NCA, now called the National Information Society Agency (NIA), to implement the

NBIS (see chapter 3). From its inception in 1986, the NCA had the central role in e-government project implementation. Projects for resident registration, real estate, automobiles, employment, customs, and economic statistics were designated as the six priority tasks, given the impact of these e-government systems on citizens (Song 2004). After these pilot projects were established in partnership with private IT companies, the government was ready to provide more integrated online public services for citizens and assess the future demand for online services. In 1993, the NCA and public administration scholars conducted a study of the future challenges of government service delivery, including an analysis of public service demand for the 2010s and of IT adoption in the public sector in the United States and Japan (Ahn et al. 1993).

The NIA, in collaboration with a research institution in Korea University, also conducted a stakeholder analysis of civil servants and citizens for developing customized e-government systems, by conducting surveys and interviews of these stakeholders and consulting IT experts on feasibility issues (Suh et al. 1996). The assessment study ultimately presented a roadmap of a citizen-centered, one-stop service e-government development strategy that emphasized the project goals of integrated e-government systems and responsiveness to citizen needs. It also indicated the types of projects, the timing for project delivery, and a cost-benefit analysis by project.

Another important project management undertaking was a pilot study for local government's e-government development through reengineering government service delivery systems (Suh et al. 1996). In 1996, the NIA conducted an in-depth pilot study of e-government development in Gangnam-gu, one of 25 local districts in the southern part of Seoul (see annex 2A). As a result of a local e-government model developed through careful reengineering analysis, Gangnam-gu became the leading district for application of an advanced one-stop service e-government system that had been carefully designed to take into consideration both internal operations for civil servants and the quality of service for citizens (Kim 2008). The National Committee of E-government Development in the Republic of Korea recognized the e-government development in Gangnam-gu as the best e-government practice in the country for three consecutive years, starting in 2001 (Kim 2008).

Financing and Human Resource Capacity Building

Finance for Digital Government Development

Implementation of e-government results in big changes in how decisions are made and how work is done. In this regard, financing and capacity-building are essential while initiating an e-government project with proper investment in human resources (Lee 2012a; SCeG 2003). It is also important to secure a medium- to long-term budget plan that covers at least three to five years, and the budget plan needs to be developed by an organization with appropriate authority to enable it to take charge of planning and managing the implementation process. Korea provides a good example of such an approach, as it successfully secured the financial resources for a five-year plan from the initial stage of e-government (Lee 2012a; SCeG 2003). Specifically, the Korean government created an information technology promotion fund and systems for pre-investment and post-settlement. This method of funding was considered revolutionary at the time. Nonetheless, it worked relatively smoothly because the government introduced a new method of investing through the subsidiaries of public institutions and recouping the costs by charging fees for use of the administrative network (Lee 2012b).

Regardless of how good a policy may be, it is useless if it is not feasible. To increase the viability of a policy, it is critical to secure the necessary budget and relevant resources. In the early stages of e-government, the Korean government had a detailed budget plan and strategy for securing the necessary financial resources for the project. One of the approaches the Korean government took was "investment first and settlement later," which was used from 1987 to 1992 (Lee 2012b).

The first comprehensive budget plan for e-government projects appeared in the Administrative Computing Network Plan that was formulated in 1987. Several budget plans had been drafted previously, but, due to lack of financial resources, the plans could not be carried out until 1987. DACOM, a company that took charge of system design and software development in the National Basic Computing Network Project, could not secure the budget, and the company's activities were implemented only several years later (Lee 2012b; MOIC 2003; NIA 2006). Moreover, intangible components of the e-government project, such as software development, were excluded from the national budget. Thus, the inability to secure the necessary financial resources at the outset was a key issue in the National Basic Computing Network Project.

To provide sustainable financial investment in the national e-government project, the Korean Telecommunication Promotion Corporation (KIPC) was created in 1986 as a 30 billion-won company (MOIC 2003; NIA 2006). As soon as KIPC was launched, it announced that a total of 151.3 billion won would be invested in DACOM, and the planned projects were carried out successfully. The fact that the government invested through a subsidiary corporation and repaid the company through user fees can be attributed to the expectation and belief that the establishment of an administrative network would create substantial value.

The general account budget and the IPF were established in 1993, during the Foundation stage. The purpose of the IPF was to financially support informatization projects, advance and develop ICT infrastructure, promote research and development in the ICT industry, and support human resource training either as an investment or a loan (figure 2.2). The IPF also encouraged the development of the required technology and the introduction of high-priced equipment by lending funds at low interest rates to small and medium-size ("venture") companies, which would otherwise not have been able to afford such equipment because of the high risks and the large scale of investment.

Figure 2.2 Annual Budget Spent on e-Government (Full Promotion Stage, 1996–2002, Unit: Korean Won 100 million)

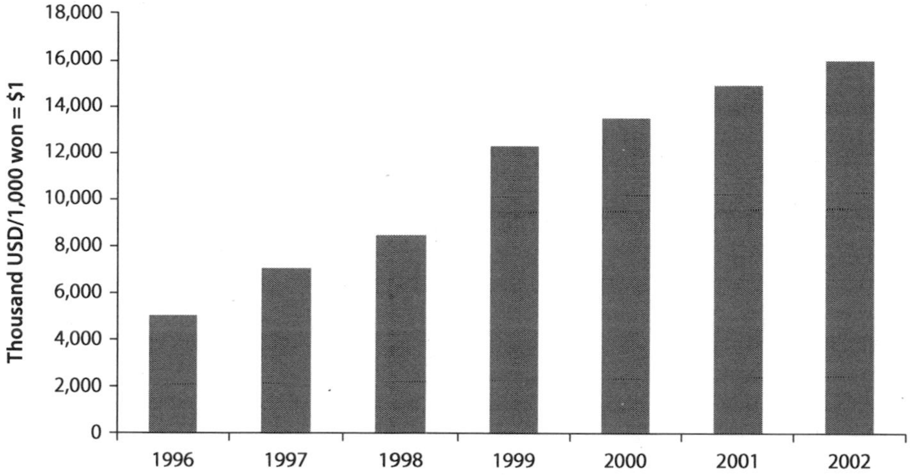

Source: Ministry of Planning and Budget, recreated from its Summary of the Budget for Fiscal Years 1996–2002.

For the first National Basic Computing Network Project, the financial resources came from the national budget account, while the IPF was used for the First E-Government Plan. The IPF was abolished during the Second E-Government Plan, and the government secured the budget from the general account. MOGAHA initially drew up the e-government budget after surveying the demand for e-government activities from individual ministries, and then implemented it after discussions with the related agencies, including the Ministry of Planning and Budget.

Human Resource Capacity: Information Technology and Digital Government Training

IT and Digital Government Training for Government Employees

One of the most important success factors of the Korean e-government development has been investment in IT and e-government education for civil servants (table 2.2). Korea's investment in IT training for government employees should be understood in the context of the civil service system, which has consistently emphasized training and career development. Since the merit-based civil service system was established in 1963, a goal of personnel management was to connect the training programs of civil servants to the public employee evaluation system. The Park administration established several training institutions and programs to enhance government employees' skills. Since the National Computer Center was established in the Science and Technology Agency in 1971, the computer and e-government training programs for government employees have been a main strategy for developing human resource capacity for e-government development in the Korean government (Park 2009). In 2015, the "smart training" division of

Table 2.2 Annual Numbers of Public Servants Who Received Informatization Training

Type of training	Full program stage	Advanced stage								
	2001	2002	2003	2004	2005	2006	2007	2008	2009	2010
Offline	10,026	8,695	6,257	6,246	6,256	5,210	4,226	3,797	4,982	14,423
Online	456	1,383	1,803	2,403	3,592	3,069	4,412	7,355	5,661	5,033
Total	10,482	10,078	8,060	8,649	9,848	8,279	8,638	11,152	10,643	19,456

Source: National Information Society Agency 2011: 295.

the Central Officials Training Institute (COTI) provided IT education for government officials, reflecting the increased demand for cyber-education and e-learning by using PCs, smart phones, and tablet PCs.

From the very early stage of e-government development in the 1980s, a pilot project, such as the computerization of post offices under the Administrative Computing Network Plan, and specific training plans were concurrently established for the successful implementation of the particular pilot project (NIA 2001). The IPF, established in January 1993, was also used to promote ICT research and development and ICT human resource training. Since 1997, around 30,000 public employees per year took IT training and other e-government training programs at the national and local government levels as well as at the agency level. Training programs were also contracted out to the private sector (NIA 2001).

Another example of IT capacity-building is the informatization of local finance in the early 2000s, which required a new online performance budgeting and accounting system at the local government level. MOGAHA selected 21 local government agencies as targeted training groups of the local finance system, and designated the local governments as the mentor agencies that taught and promoted the new system to the other local government civil servants (Lee 2012b). Today, the Korea Local Information Research and Development Institute, established in 2008 through the collaborating efforts of 16 metropolitan city and provincial governments, supports the effective operation of the local finance system in local governments.

To enhance the competence of e-government management and information policy formulation and implementation, more training programs targeting senior-level managers were also offered beginning in 2006. For senior-level managers, during the fully developed stage of e-government between 2003 and 2012, the government offered customized training programs on the integrated On-nara online system (Park 2009). Senior-level civil servants and political appointees, including the ministers and vice ministers, received similar training (Lee 2012b).

IT Education for Citizens
Another important human resource capacity-building strategy for e-government development was investing in IT education for citizens. In 1997, there were

1.9 million Internet users in Korea, and that number grew to 26.27 million by the end of 2002. The driving factor for the increased use of the Internet by Koreans is directly related to government investment in IT education for citizens, especially under President Kim Dae-jung's commitment to e-government development (1998–2002). To minimize public concerns about the digital divide, the government paid special attention to free IT education for those who might be isolated from the information age, including homemakers, soldiers, seniors, people with disabilities, and even prisoners (Ko and Kang 2014; MIC 2003). From 2000 to 2002, a national campaign for "IT Education for 10 Million Citizens" was launched under the leadership of MIC and nine other ministries. Table 2.3 compares the targeted and actual number of beneficiaries, while table 2.4 outlines the content of the HRD project for the IT sector.

The targeted groups for IT education included housewives, local residents, white- and blue-collar workers, and students. The government also set a training target for people with disabilities, farmers, fishermen, senior citizens, inmates, juveniles, soldiers, the underprivileged, government employees, teachers, and employees of public corporations. The training was delivered by ICT education facilities in the public and private sectors, and the fees were subsidized to enable people to access computers and the Internet (Ko and Kang 2014). The Internet training was delivered to students and teachers in public schools, and a training certificate was issued for high school students from 1999 and for junior high school students from 2001 to stimulate ICT education. ICT training programs for teachers and school administrators were expanded to 37 percent (125,398)

Table 2.3 Accomplishments of the IT Education for 10 Million Citizens Project
(Unit: 1,000)

Education target	Targeted beneficiaries	Actual beneficiaries
Disabled person	206	101
Farmer	171	129
Fisherman	20	16
Senior citizen	171	443
Housewife	2,000	434
Inmate	32	120
Local resident	1,600	5,359
Worker	1,500	1,435
Soldier	740	623
Civil servants	510	510
Teacher	615	1,109
Student	3,364	3,373
Employee and executive of public corporation	200	153
Total	**11,130**	**13,850**

Source: Ko and Kang 2014, 34.
Note: ICT = information and communication technology.

Table 2.4 Content of Human Resource Development Project

Classification	Project details	Budget
Support for ICT-skilled, gifted young talents	Host a computer creativity contest to discover gifted young ICT talent; provide further ICT education for young talents	700 million won
Informatization education support for the disabled	Provide informatization education for the disabled to promote social participation via the Internet and to enhance their ability to adapt to the information age	2 billion won
Informatization education support for females	Provide informatization education for females to promote the social participation of the female labor force	2 billion won
	Support the special education institutions for women operated by Gifted Scout of Korea and Central Women's Association	
Informatization education support for soldiers	Provide information and communication education to young soldiers to equip them with digital knowledge required for modern warfare	2 billion won

Source: Ko and Kang 2014, 65.
Note: ICT = information and communication technology.

of total teachers and administrators (MIC 2003). In 1999, the government also provided PCs at a low price to low-income people and residents in rural areas (MIC 2003).

ICT Human Resources Development

One of the success factors of Korean e-government development is the long-term proactive investment in ICT human resource development through education and training opportunities. The government developed and implemented the ICT HRD policies in response to changes in domestic and foreign circumstances. The training and education for ICT HRD has been carried out through various channels, such as formal education institutions, private training centers, learning centers run by public institutions, education institutions abroad, and corporations (Ko and Kang 2014).

The ICT policies were first introduced in the 1970s and were included in the First Five-Year Basic Plan for Administrative Computerization (1978–82). However, the policies for ICT HRD began only in the early 1990s when the National Policy Plan for Information Industry Promotion (1992) was established, followed by the First National Informatization Promotion Master Plan (1996–98). In the 1990s, technologies were changing rapidly, but the necessary human resources were lacking, especially in the areas of software and information and communication. High-quality human resources with master's or doctoral degrees were in short supply, and basic-level human resources were oversupplied. The education institutions had not properly responded to the situation. The government realized there was a serious problem and began to take measures to train and educate citizens and a potential IT workforce. The roles and responsibilities for HRD were shared by several ministries: Ministry of

Culture and Education (MOCE), Ministry of Commerce and Industry (MOCI), Ministry of Post and Telecommunications (MOPT), and Ministry of Science and Technology (MST).

In the years following the Asian Financial Crisis in 1997, Korea experienced mass unemployment and faced a growing need to retrain workers. Soon after the crisis, a software venture boom started, and, from the 2000s onwards, conglomerates such as Samsung began to expand their global market share in the ICT industry, which, in turn, fueled the industry's demand for a larger workforce. In addition, as the high-speed Internet was introduced and proliferated throughout the nation, interest in informatization grew among the general public. These rapid changes produced a high demand for human resources in the ICT industry and the need to retrain the unemployed. In response, the ICT HRD policies moved forward with a focus on building infrastructure to support the expansion of ICT HR (Ko and Kang 2014).

However, firms were complaining that IT people produced by the education institutions lacked the knowledge and skills required in business (MOE 2003). Such situations led the ICT HRD in 2004 to focus on the following: cultivating demand-oriented human resources; training advanced-skilled ICT professionals with global competitiveness; and enhancing ICT-specialized education for the industrial workforce (MOE 2003; NIA 2004, 2008). Aiming to produce human resources with the capacity to develop core ICT technologies as well as to enhance the R&D capacity of existing universities, the government designated universities with high quality in education and research as Information Technology Research Centers (ITRCs) and provided them with substantial support. As a result, the number of ITRCs increased from 32 in 2002 to 50 in 2007, while producing around 1,000 graduates with master's or doctoral degrees. To enhance the capacity of the industrial workforce, mid- and long-term retraining programs were provided to existing ICT workers (NIA 2008).

As a consequence, the average career spans of existing ICT workers were extended from 6.9 years in 2003 to 8.4 years in 2005 and 11.1 years in 2008 (Ko and Kang 2014). With the launch of the Lee Myung-bak administration in 2008, the focus of the ICT HRD policies moved toward cultivating convergence and software specialists. Korea signed the Seoul Accord in December 2008 in an attempt to take the lead in international standardization of engineering education and facilitate personnel exchanges in ICT fields with other countries (Ko and Kang 2014; NIA 2013).

According to Ko and Kang, the objectives of Korea's ICT HRD policies have progressed over time to (1) establish an ICT foundation; (2) expand the ICT HR pool through increasing enrollment quotas in ICT-related programs in education institutions; and (3) improve the quality of ICT education by enhancing linkages between education institutions and ICT firms (Ko and Kang 2014). In sum, strategic and systematic targets and plans for government-led basic ICT education in Korea were not only beneficial for e-government development, but also helped citizens to acquire basic ICT skills.

Bringing Government into the 21st Century • http://dx.doi.org/10.1596/978-1-4648-0881-4

Conclusion

The path taken by Korea in its e-government development shows the importance of three critical factors:

- *Establishing institutions and laws at an early stage.* This made it possible to coordinate the various ministries involved, and secure the necessary budget in advance for IT infrastructure building;
- *Having long-term presidential leadership.* The greatest driving force in pursuit of e-government development in Korea for more than 30 years has been the leadership role of the various presidents. They have all had a clear vision of e-government development, a strong commitment to delivery, and leadership skills for coordination; and
- *Putting appropriate mechanisms in place.* Korea's e-government development has also been driven by mechanisms for designing, delivering, and tracking activities and performance relative to the overall e-government objectives.

Government officials in other countries, who plan to connect the development of e-government to government reform, can usefully take note of the Korean government's strategy for its e-government transformation and the lessons learned.

1. *Secure financial resources.* Leaders in other countries should keep in mind that a national vision, development strategy, and contingency plan should be established and implemented systematically with secured financial resources and the help of the right experts. During the groundwork stage of e-government development, the scope and level of the project expands beyond individual organizations, and reaches the national level. These challenges should be addressed in the early stages with feasible financial resources.

2. *Focus on improved services.* During the integration stage, the focus should be on identifying how e-government will be used to improve government administration and services, such as public services for citizens, businesses, and other public institutions. Standards for government administration and services should be continuously assessed to improve public services, enhance government efficiency, and promote accountability by establishing institutions and enacting relevant laws to meet new challenges in the era of digital governance.

3. *Ensure collaboration and coordination.* The SCeG shows that if leaders are committed to improving integrated services through e-government, they will encourage collaboration among the parties concerned as well as ongoing feedback on e-project management. Also, efforts must be made to strengthen coordination so as to address any conflicts among various stakeholders. As public institutions are more integrated under e-government, cooperation becomes

more important than ever. When conflict arises among public institutions, leadership can play a major role in resolving these conflicts.

4. *Demonstrate the success of pilot e-government projects.* Demonstrating successes is important to attract more resources and provide more motivation to push on with the e-government project. In Korea, completion of the 11 e-government projects from 2001 and 2002 provided such motivation.

5. *Build operational leadership capacity:* Based on the lessons from the pilot projects, the Korean government noted the importance of operational leadership for effective delivery of the e-government project, including the forecasting analysis of the demand for future ICT services, stakeholder analysis, feasibility assessment, and a roadmap for prioritizing short-term and long-term projects.

6. *Invest in human resource development.* One of the most important capacity-building strategies for e-government development in Korea is the government's long-term investment in HRD. The government provided long-term IT training not only for government officials but also for every citizen, as an attempt to overcome the digital divide based on region, gender, age, and profession.

Other lessons can be learned from the weaknesses in Korea's pursuit of e-government development. These weaknesses include:

- *The lack of leadership attention in the early stages to institutionalizing cost-efficiency and cost-benefit analyses of e-government projects.* Insufficient time was given for the preparation of critical assessment studies or for the creation of organizational learning systems using information on e-government successes or failures. Such analyses would have facilitated changes or modifications in e-government goals, application systems, and services.
- *The limited investment in a bottom-up approach.* Such an approach allows core users' input and ideas to be collected and discussed before the design of any e-government project or service. This related, in particular, to concerns about security and privacy issues among stakeholders affected by the e-government projects. Korean e-government development can be assessed as very much a top-down approach, similar to the success of Korea's economic development between the 1960s and the 1970s.

In sum, presidential leadership has been the most significant success factor of the Korean e-government development. Such leadership has enabled the integration of a clear e-government vision and goals, effective communication, and the required resources. Successful e-government innovation relies on the development of strong management capacity to lead taskforce teams involving public–private partnerships (such as the national e-government committee), operational leadership and collaboration at the project management level, human resource management capacity for ICT, and IT education for citizens and government employees.

Annex 2A: Case Study of Gangnam-gu on Collaboration: The Pilot Project for Local Digital Government Development

Based on the pilot study in Gangnam-gu, an elected district mayor was motivated to transform the local government to be a leading e-government district with high-quality online services, transparency, and online citizen participation. To establish e-government innovations and e-participation, the mayor institutionalized several dimensions of management capacity between 1997 and 2003. This annex summarizes the capacity-building strategy of Gangnam-gu's e-government innovation, including human capital, e-citizenship-building efforts, and IT capacity and outsourcing (Bretschneider et al. 2005; Kim 2008). The Gangnam-gu case demonstrates that the executive leader's capacity to integrate clear e-government vision and goals, effective communication, and appropriate management systems is vital to successful e-government innovation at the local level (Kim 2008).

Decentralization and leadership of the elected local mayor. An in-depth case study on Gangnam-gu e-government was conducted in 2003 (Kim 2008; Kim, Lee, and Kim 2008). It showed the district mayor's leadership as a leading driving force of e-government initiatives and development. Among 45 interviews of public employees in the district conducted in 2003, 23 interviewees indicated the mayor's leadership as an enabler of e-government initiatives and development. Five interviewees also pointed out division directors' and the vice mayor's leadership as a factor affecting the success of e-government in Gangnam-gu.

Leadership with clear vision and goals is an essential ingredient for successful adaptation and management of technology by government, and executive leaders should prepare effective plans and targets that provide a roadmap for the future of e-government development in the context of the newly established decentralized local government system in Korea. In Gangnam-gu, the mayor achieved these dimensions of leadership, as Gangnam-gu managers state:

> The mayor always emphasizes that we should be the best in the world in the IT aspect. It is the will and intention of the head of an organization.

> He has achieved that kind of goal and in our view, our mayor is an excellent administrator and his vision is 10 years ahead of us and we are always busy following it. His vision is way ahead of us and considering this, it is amazing he got us this far.

Communication. One of the early and continuing initiatives of the mayor of Gangnam-gu during his first term was to promote government transparency for citizens, via an electronic and web-based system, known as the e-democracy type applications. Weekly cabinet meetings were broadcast over the web, a form of one-way communication from government to citizen, while e-mail and bulletin board-type services provided two-way communications (Bretschneider et al. 2005).

Investment in HRD. Gangnam-gu provided free IT education for citizens to improve their usage of online services and e-participation. The classes were held in 18 elementary and junior high school facilities and eight district offices; 20,550 citizens received IT education in 2002, and 23,718 in 2003. Gangnam-gu also paid attention to marketing IT education (Bretschneider et al. 2005). The mayor supported employee training programs in response to changes in the organizational environment. In 2002, Gangnam-gu established the Gangnam-gu Academy Hall, where all internal employee training programs were held. Various training programs were developed, including local autonomy rules and guidelines, contract management, customer management, language skills, research surveys, and management. In 2002, 80 classes were provided and 2,005 employees received training. Around 2,320 employees took IT and e-government-related courses in 2003. Gangnam-gu also developed e-learning systems to increase the flexibility of the training programs for employees (Bretschneider et al. 2005). After establishing a partnership with several private corporations in 2002, Gangnam-gu developed a two-week intensive management training program for employees and also provided them with continuous training and education, via a formal partnership with the Seoul City University in 2005 (Bretschneider et al. 2005). The investment in HRD showed the mayor's commitment to institutionalizing continuous training for local government employees and citizens.

Collaborative leadership: The Gangnam-gu case also demonstrates that local e-government leadership faces continuous but varying degrees of challenges related to intergovernmental collaboration, interagency coordination, and inter-sectoral partnership building (Bretschneider et al. 2005; Kim, Lee, and Kim 2008). At the initial stage of local e-government development, Gangnam-gu paid attention to building a collaborative relationship with authorizing external government organizations to acquire legal and institutional endorsement for e-government transformation, and also financial resources from upper-level governments (e.g., national government agencies and Seoul Metropolitan Government). Furthermore, during the e-government development stage, building a business relationship with private vendors became a primary collaboration task for Gangnam-gu. At the final stage of integrating e-government applications, the demand for interagency and inter-sectoral collaboration increased to its peak as the local government sought to integrate e-government applications horizontally (Kim, Lee, and Kim 2008). Kim et al. note that collaboration between upper- and lower-level governments is especially important when the lower-level government makes significant early progress in e-government transformation. The experimental approach of the local government has a chance to influence upper-level government's policy decisions for vertical integration by setting the standards for integration. Without a centralized formal system for guiding intergovernmental collaboration, each e-government project team leader at the Gangnam-gu made a

Bringing Government into the 21st Century • http://dx.doi.org/10.1596/978-1-4648-0881-4

commitment to communicate effectively and continuously with agencies at different levels of government through the creation of a Special IT Project Committee (Kim, Lee, and Kim 2008).

In conclusion, the Gangnam-gu case provides several lessons for project management and collaboration for e-government development at the local government level. The most important lesson is to create a culture of ownership where employees share information about e-government initiatives, collaboration, and best practices through ongoing project updates with staff, division leaders, department heads, and officials (Kim 2008; Kim, Lee, and Kim 2008). The second lesson for e-government leadership at the local level is to create a culture of organizational learning by encouraging employees to analyze past e-government collaboration attempts, both successful and failed, and to suggest how the organization can apply those lessons to further improve collaboration practices. Finally, the empowerment of IT project managers, and senior- and middle-level managers could facilitate intergovernmental, interagency, and inter-sectoral collaboration in achieving higher e-government performance and organizational effectiveness.

Bibliography

Ahn, M. S. 2008. *The Theory of E-Government in Korea*. Seoul: Bak-Yeong Press.

Ahn, M. S., J. S. Kim, J. J. Lee, D. H. Kim, G. S. Yoon, I. J. Jung, and S. J. Park. 1993. *A Study of the Development of Next Generation Computing Network Service for Responding to Changes in Administrative Environments in the 21st Century*. Seoul: Institute for Administrative Issues in Korea University.

Bretschneider, S., J. Gant, S. Kim, H. Choi, H. Kim, M. Ahn, and J. Lee. 2005. *E-Government in Gangnam District: Evaluating Critical Success Factors*. Center for Technology and Information Policy, Maxwell School, Syracuse University, Syracuse, NY. Project report submitted to Gangnam-gu.

Chen, Y., and D. Dimitrova. 2008. "Civic Engagement via E-Government Portals: Information, Transactions, and Policy Making." In *E-Government Research: Policy and Management*, edited by D. Norris. Hershey, PA: IGI Publishing.

Chung, C. S. 2012. *The Theory of Electronic Government*. 3rd. ed. Seoul: Seoul Economy and Business Management.

Fountain, J. E. 2004. *Building the Virtual State: Information Technology and Institutional Change*. Washington, DC: Brookings Institution Press.

Garson, D. 2006. *Public Information Technology and E-Government: Managing the Virtual State*. Sudbury, MA: Jones and Bartlett Publishers.

Heeks, R. 2001. *Reinventing Government in the Information Age: International Practice in IT-Enabled Public Sector Reform*. London: Routledge.

Hwang, J. S. 2000. "A Comparative Analysis on the E-Government's Vision and Strategy in Major Leading Countries." Seoul: National Information Society Agency.

Ingraham, P. W., J. E. Sowa, and D. P. Moynihan. 2004. "Linking Dimensions of Public Sector Leadership to Performance." In *The Art of Governance: Analyzing Management and Administration*, edited by P. W. Ingraham and L. E. Lynn, Jr., 152–70. Washington, DC: Georgetown University Press.

Killian, W., and M. Wind. 1998. "Changes in Interorganizational Coordination and Cooperation." In *Public Administration in an Information Age: A Handbook,* edited by I. Th. M. Snellen and W. B. H. J. van de Donk, 273–91. Amsterdam: IOS Press.

Kim, D. J. 2010. *Kim Dae-Jung's Autobiography.* Seoul: Sam-In Press.

Kim, H. J., J. Lee, and S. Kim. 2008. "Linking Local E-Government Development Stages to Collaboration Strategy." *International Journal of Electronic Government Research* 4 (3): 36–56.

Kim, S. 2008. "Local Electronic Government Leadership and Innovation: South Korean Experience." *Asia Pacific Journal of Public Administration* 30 (2): 165–92.

Kim, S. J. 2011. "The Success Factors and Future Tasks of the Korean e-Government." *Korean Balanced Development Studies* 2 (3): 55.

Ko, S. W., and H. Y. Kang. 2014. "2013 Modularization of Korea's Development Experience: ICT Human Resource Development Policy." Seoul: Ministry of Strategy and Finance (MOSF), KDI School of Public Policy and Management.

Kwon, G. H. 2011. *The Theory of E-Government.* Seoul: Pakyoungsa. (in p. 39)

Lee, J. H. 2012. *A Study on the Restructuring of Government Organization for E-Government.* Research Series 2012–06, Seoul: Korea Institute of Public Administration.

Lee, Y. B. 2012. *2011 Modularization of Korea's Development Experience: The Introduction of E-Government in Korea.* Seoul: Ministry of Strategy and Finance (MOSF), KDI School of Public Policy and Management.

Millard, J. 2008. "E-Government for an Inclusive Society: How Different Citizen Groups Use E-Government Services in Europe." In *E-Government Research: Policy and Management,* edited by D. Norris. Hershey, PA: IGI Publishing.

Ministry of Security and Public Administration. 2014. *Government 3.0 Development Plan Report.* Committee of Government 3.0. Seoul: Republic of Korea.

MOE (Ministry of Education and Human Resources Development). 2003. "Action Plan: Master Plan for Human Resources Development in National Strategic Areas (6T)." http://www.nhrd.net/board/view.do?boardId=BBS_0000004&menuCd=DOM _000000102003000000&orderBy=register_dt%20DESC&startPage=365&data Sid=14271.

MOIC (Ministry of Information and Communication). 2003. "Strategy for Informatization of Korea." Seoul: Ministry of Information and Communication, Republic of Korea.

Moore, M. H. 1995. *Creating Public Value: Strategic Management in Government.* Cambridge, MA: Harvard University Press.

NIA (National Information Society Agency). 1994. "1994 Informatization White Paper." http://www.nia.or.kr/.

———. 1996. "1996 Informatization White Paper." http://www.nia.or.kr/.

———. 2001. "2001 Informatization White Paper." http://www.nia.or.kr/.

———. 2004. "2004 Informatization White Paper." http://www.nia.or.kr/.

———. 2005. *The History of E-Government Policy Development in Korea.* Seoul: National Information Society Agency.

———. 2006. *20 Years History Book of NIA.* Korean Publication.

———. 2008. "2008 Informatization White Paper." http://www.nia.or.kr/.

———. 2011. "2011 Informatization White Paper." http://www.nia.or.kr/.

———. 2013. "2013 Informatization White Paper." http://www.nia.or.kr/.

Norris, D. 2008. *E-Government Research: Policy and Management*. Hershey, PA: IGI Publishing.

OECD (Organization for Economic Co-operation and Development). 2003. *Checklist for E-Government Leaders*. Paris: OECD, Public Affairs Division, Public Affairs and Communications Directorate.

O' Looney, J. 2002. *Wiring Governments: Challenges and Possibilities for Public Managers*. Westport, CT: Greenwood Publishing Group.

Otenyo, E., and N. Lind. 2011. *E-Government*. Amherst, NY: Teneo Press.

Park, C. S. 2009. "Policy Direction for Development of Information Capacity of the Public Officers." *Local Government Review* 15 (3): 40–45.

SCeG (Special Committee for e-Government). 2003. "2003 White Paper of E-Government." Seoul: Special Committee for E-Government.

Soh, Y. J. 2003. "Overcoming the Dilemma in the Structure of E-Government Building Project: In the Case of E-Government Special Committee." *Informatization Policy* 10 (2): 30–49.

Song, H. J. 2004. *Building E-Governance through Reform*. Vol. 2. Seoul: Ewha Womens University Press.

Song, H. J., and T. Cho. 2007. "Electronic Government of Korea: Performance and Tasks." *Informatization Policy* 14 (4): 20–37.

Suh, S. Y., K. H. Jung, Y. S. Kim, K. S. Oh, G. S. Yoon, and B. G. Jung. 1996. "A Study of the Administrative Reform in Local Government." (NCA V-PER_95123). Seoul: National Information Society Agency.

Sung, D., and C. Jang. 2005. "Evaluation of Customer-Orientedness of Public Services: Focused on the e-Government. *Korean Public Administration Review* 39 (2): 207–32.

Tichy, N. M., and M. A. Devanna. 1986. *The Transformational Leader*. New York: John Wiley and Sons.

Titah, R., and H. Barki. 2008. "E-Government Adoption and Acceptance: A Literature Review and Research Framework." In *E-Government Research: Policy and Management*, edited by D. Norris. Hershey, PA: IGI Publishing.

Yoon, Y. M. 2003. "Leadership and Coordination of the E-Government: Lessons from the People's Government." *Quarterly Thought* 57 (Summer): 55–74.

CHAPTER 3

Korean Digital Government Infrastructure Building and Implementation: Capacity Dimensions

Jeongwon Yoon

Introduction

The Republic of Korea has taken a radical and strategic approach to develop its infrastructure and digital-driven economy. Its decision to leapfrog from a home appliance manufacturer to a player in information and communication technology (ICT) stems from the notion that ICT infrastructure would eventually be the foundation of a competitive public service sector and economy.

One of Korea's most innovative strategies was the building of the world's fastest telecommunications network. Along with the development and growth of computer technology and the Internet, implementation of infrastructure that allows rapid transmission of information greatly enhanced Korea's competitiveness. With its information and communications infrastructure, Korea was able to provide world class e-government public services and develop the ICT industry and new technology to support it. As a result, Korea now possesses the most advanced and fastest ICT infrastructure, and is earning more than 25 percent of its gross domestic product (GDP) from the ICT and related service sectors. Ultimately, such an infrastructure will lead to the growth of an environment that delivers diverse types of services and the growth of relevant industries, which will, in turn, lead to more investment and the hiring of high-quality professional manpower, thereby creating an ongoing virtuous cycle.

Various ICT-related indexes assessed by international organizations show Korea's growth in recent years. The Organisation for Economic Co-operation and Development (OECD), ITU, and the UN describe Korea as a country with a highly developed ICT infrastructure or the most advanced e-government service delivery. Korea's reputation is drawing attention and pushing the per capita GDP to almost US$30,000 a year, helping Korea become the world's 13th largest economy. Of the US$500 billion worth of goods Korea exports globally each year, ICT exports run to about US$170 billion. With such economic growth, the billions invested in the information and communications infrastructure do not seem too large an investment. Many countries are starting to understand the importance of ICT infrastructure to foster growth (Yoon and Chae 2009), but many countries also have questions about how Korea accomplished its ICT development and fast adaptation:

1. How did Korea realize early on that developing technology and investing in ICT infrastructure would be beneficial for the delivery of public services (e-government)?
2. How did Korea finance the needed funds for the investment?
3. How was the national strategy for investment and implementation developed, and what were the factors that directly affected the decision-making process for the e-government projects?
4. What were the challenges in the implementation of advanced e-government infrastructure, and how were these difficulties overcome?
5. What kind of economic benefits did building such public infrastructure bring about for Korea?
6. How did the Korean government use the implemented infrastructure efficiently, and what were the services that were provided to enhance work efficiency in the government?

To address these questions, this chapter will review the origin of Korea's strategy to build the values and systems of digital governance through the development of ICT infrastructure, discuss how Korea moved forward in building an increasingly advanced ICT infrastructure, and finally consider Korea's future direction and strategy to maintain its position as a leader in e-government. Clearly, there were challenges along the way, and so the chapter also draws lessons from both the successes and the failures.

Stepping Stones in the 1970s and 1980s

To understand the motivation for the government's initiation of large-scale high-speed network projects, as well as the success factors that led to Korea's growth into an ICT industry-oriented nation, we need to look briefly at Korea's telecommunications infrastructure and industry in the 1970s and 1980s. After the destruction caused by the Korean War, beginning in the 1970s, Korea put a lot of effort into building a strong industrial infrastructure by fostering and promoting industries such as oil refining, shipbuilding, steel and automobile manufacturing, and construction. Korea was able to benefit from large-scale, government-led national projects. The country also enjoyed growth from an export-oriented economic development policy to overcome its limited population and land size and the lack of natural resources. At the end of the 1970s, when Korea began to prepare to host the Summer Olympic Games in 1988, it recognized that the telecommunications infrastructure was just as important as the industrial infrastructure and that the electronics industry has a great impact on the telecommunications industry (Larson and Park 1993). The realization that the advantages enjoyed by developed nations came from highly developed and advanced telecommunications networks prompted Korea to become more competitive by developing its own advanced telecommunications network in the early 1980s.

Finding that an expensive electronic switching device was one of the main obstacles against expansion of the telecommunications network, Korea decided to develop the necessary technology to localize production of the device. At the time, only six developed nations (Belgium, France, Japan, Sweden, United Kingdom, and the United States) possessed the technology to manufacture the electronic switching device. Moreover, without the electronic switching device, the mass introduction of the landline telephone in Korea was progressing too slowly, with installation in the average home being almost impossible at that time. It became obvious that local development of such technology was a high-risk idea. Nevertheless, Korea decided to invest 0.9 percent of its annual GDP income into developing the device. Since this project was Korea's largest research and development (R&D) project undertaken since 1945 (Oh and Larson 2011), it would need all of the human resource (HR) and knowledge capacity accumulated as an industrializing and developing country. In the early 1980s, US$60 million was appropriated as the R&D budget for this project, and an R&D consortium made up of government-funded research institutes, academia, and private companies from the electronics industry was set up to carry out the project. With the participation of many Korean scientists and engineers living overseas, Korea succeeded in developing the electronic switching device and greatly lowered the implementation cost of these devices. This allowed Korea Telecom (KT) to deploy the domestically manufactured electronic switching device on a massive scale. With the development of the TDX-1[1] in 1985, Korea became the 10th country to possess the electronic switching device. In 1987, only two years after the development of the TDX-1, the telephone penetration rate of Korea reached 100 percent compared with a rate of only 35 percent for households in 1980. The success of the largest R&D project carried out by Korea enhanced its national brand value, had a great impact on the electronics industry, and was the true beginning of Korea's ICT industry.

After the first deployment of the telecommunications network, the Korean government started to develop midsize computers—or what were formerly called minicomputers—by using the know-how and human resources that led to the development of the electronic switching device. In particular, the midsize computer, TICOM,[2] developed by using the open source unix operating system, was distributed to local governments from the early 1990s and used to provide public services to citizens.

In essence, these events can be described as the beginning of e-government in Korea. But what was the motivation behind the risky investment of 0.9 percent of annual GDP to develop the electronic switching device?

The cost of *importing* the electronic switching device to achieve a telephone penetration rate of 100 percent in Korean households was estimated to be US$600 million to US$700 million (NIA and Ministry of Public Administration and Security 2008). Because Korea Telecom was owned by the government at that time, the cost of such an import would be a huge financial burden to the Korean government. However, developing their own device would require an investment of only one-tenth of the cost of the imported product, and if it were successful, a huge budget saving would be possible for the government. At the same time, operational costs, such as the cost of replacement parts, could be greatly reduced; highly valuable human resources in R&D development could be secured; and the foundation in the telecommunications manufacturing and service industries could be built.

Ultimately, the development of the electronic switching device invigorated the Korean telecommunications market and laid the foundation on which Korea became an advanced ICT-oriented country. Moreover, the transition of the industrial portfolio from manufacturing televisions and radios to high value-added telecommunications and service industries allowed Korea to place itself one step ahead of other developing nations. Through the implementation and development of the electronic switching device, the burden of communication costs for Koreans was greatly reduced. The low-cost telecommunications network deployed nationwide secured the foundation and system on which Korea could compete with developed nations. This element was the most-needed infrastructure for Korea because Korea was to host the 1986 Asian Games and the 1988 Summer Olympic Games. With confidence, knowledge, and capacity coming from the successful development of the electronic switching device, Korea went on to implement the National Basic Information System (NBIS) project in 1987 (Yoon 2006).[3]

The promotion of the electronics industry in the 1970s and the establishment of a basic national strategy to foster the telecommunications industry in the 1980s led to efforts to secure needed financing, hire capable technical workers, and develop the telecommunication service market. These would become the foundation for carrying out the high-speed network implementation project in the 1990s. With both preemptive and strategic planning, the innovative method of data communication using computers became possible in Korea. This was the beginning of Korea's miraculous economic rise.

Early Stage of Digital Government Infrastructure

With the widespread use of PCs and the 100 percent household telephone penetration rate as a result of the home-grown development of TDX (the electronic switching device), Korea began to realize the importance of computer networks. Korea now had a very solid foundation in telecommunications for a developing country. However, even such a telecommunications network was not sufficient to prepare for the next century. The advent of the Internet and the emergence of the Apple computer and the IBM PC between 1970 and 1990 heralded an age when computers would be used for communication.

In 1987, the Korean government began the NBIS project. The goal was to develop the network infrastructure that used computers as the means for data communication. Knowledge was acquired through developing the TDX, gaining experience in computer design and manufacturing technology, and deploying telecommunications networks. In addition, companies such as Samsung and LG gained a sound knowledge of production technology for personal computers. All this contributed to Korea's capacity to build the data network infrastructure.

In 1986, Korea passed the Act on Expanded Deployment and Promotion of the Digital Network, and established the National Computerization Agency (NCA)—now called the National Information Society Agency (NIA)—as a specialized government agency to develop the standards for the implementation and operation of the NBIS and to supervise the project (NIA 2005b). The NBIS consisted of five digital networks:

1. the administrative network for the government and government-funded institutions;
2. the financial network for banks, insurance companies, and the securities commission;
3. the education and R&D network for universities and research institutes;
4. the national defense network for defense-related organizations; and
5. the public security network for security-related organizations.

Of the five digital networks, it was the administrative network that formed the core of the NBIS project. Paper records for six types of government-related work, such as resident registration management, real estate management, employment management, customs clearance management, economic statistics, and automobile registration management, were digitized into databases for online public service delivery, and became the beginning of the Korean e-government.

Along with four private companies (Samsung Electronics, LG Electronics, Hyundai Electronics, and Trigem Computer), the ETRI (Electronics and Telecommunications Research Institute),[4] a government-funded telecommunication research institute, jointly developed a domestic midsize computer (TICOM). The Korean government then supported the development of domestic technology and helped the growth of the corresponding high-tech market indirectly by purchasing TICOM and deploying it in central government agencies and local government data centers, thereby contributing to the implementation of the digital network. This was one of the earliest examples of a joint public–private partnership (PPP) that laid the foundations for e-government in Korea. The NBIS project that began in 1987 was completed in 1996.

History of Funding and Strategic Approaches for Digital Governance

In 1993 in the United States, the Clinton administration presented a blueprint called the "Information Superhighway." The Clinton administration also announced that in the 21st century data transmission at extremely high speeds would bring about an information revolution with radical results that could critically affect the very survivability of a nation. In response, in 1995 Korea reorganized the Ministry of Communications into the Ministry of Information and Communications (MIC) and established a high-speed broadband network implementation plan that would invest US$45 billion until 2015. In 1997, MIC revised this plan to complete the implementation of the network by 2005 with a total investment of US$32 billion[5] based on the certainty that this high-speed broadband network would advance Korea's place in the world rankings of knowledge-based countries.

The government was well aware that the national digital network in use did not have the capacity to handle the vast amount of multimedia data that would be transmitted in the 21st century, and that Korea's future therefore depended on the implementation of an advanced information infrastructure. As Korea had experienced the impacts on economic growth of building national infrastructure projects—highways, ports, airports, and new cities—the government understood the importance of building the information infrastructure.

Korea also realized specifically that the sharing of information was an essential element needed to raise competitiveness and citizens' quality of life and level of education, together with the rapid democratization process that had put an end to 26 years of military dictatorship in 1987 and introduced direct presidential elections. To realize this vision, the Korean government started to establish various national informatization strategies on a regular basis. Following the high-speed broadband network implementation plan, called the Korea Information Infrastructure (KII)[6] initiative in 1995 (NIA 2005a), the Korean government passed the Framework Act on National Informatization Promotion in June 1996 and selected 11 essential high-impact national tasks needed in the information society. These tasks were completed in 2000.

But Korea was also facing the possibility of a foreign debt moratorium caused by the 1997 Asian Financial Crisis and had to comply with corporate restructuring as mandated by the International Monetary Fund (IMF) as a condition for receiving a financial bailout. Many companies went bankrupt and the manufacturing-oriented Korean economy suffered significantly. Even in this difficult situation, firmly believing that revitalizing the ICT sector would help to revive the economy, Korea launched the Cyber Korea 21 initiative in 1999 as the strategy for national informatization. The main goal of Cyber Korea 21 was the creation of new jobs through informatization. Many start-up companies were created, taking advantage of newly formed large-scale IT projects, diverse support programs, and market formation of informatization services offered by this initiative. The dotcom bubble had a considerable negative impact on the huge number of ICT companies that had been created, and many of them did not survive for long. However, the Cyber Korea 21 initiative paved the way for future directions in using ICT to overcome the economic crisis. It also provided ICT companies with a platform for growth, in tandem with the KII project. In addition, based on this initiative, diverse campaigns and educational programs to reduce the digital divide for Korean citizens were carried out. It was at this time that the words "knowledge management" were first introduced.

In 2002, Korea announced the e-Korea Vision 2006. This initiative was focused on public service delivery by making full use of the high-speed broadband infrastructure that had been widely implemented in Korea by then. To provide e-government services based on the KII to the people and public institutions, the government selected an additional 11 major e-government initiatives and invested several hundred billion Korean won each and every year. As these initiatives began their service delivery, Korea's online public service delivery that made use of the KII infrastructure started to take shape. The words *digital economy* first appeared in the Broadband IT Korea Vision 2007. The Korean government firmly believed that the digital economy would become a pillar of the economy and thus it was a major part of its strategic plan. This was made clear through the BcN (Broadband Convergence Network) implementation plan (NIA 2006a) that integrated the e-government roadmap with audio, broadcasting, and communications.

Implementation of Digital Government Architecture

Korea Information Infrastructure

Korea went through many reviews, discussions, and evaluations when it decided to invest in the implementation of the KII. The KII was divided into three parts for roles and allocation of funds. The KII-Government was to be used by the government/public sector; KII-Public was for the private sector (home and business users); and KII-Testbed was for education and research institutes.

- *KII-Government.* The government installed a high-speed broadband infrastructure implementation commission at the working level to determine the size of funding that would be needed and the method of implementation for deploying the KII-Government. Participating organizations within this working-level commission included public agencies—MIC, NIA, ETRI—and private agencies—KT and Dacom—who jointly announced a basic implementation plan in August 1993 after extensive discussions with MIC and other relevant government agencies. The government made a direct investment of US$1 billion for KII-Government, which was used to settle the fee for its use.

- *KII-Public.* In accordance with the Framework Act on Telecommunications, KII-Public was to be deployed by making use of telecommunications facilities and networks possessed by communication business operators. The government would lease or sign a contract (permanent, 10 years, or 20 years) for its use and make partial payments for the implementation of the network and facilities to own it and to support the cost of its operation. The Board of Audit and Inspection revised this plan so that the same communications business operators for implementation and deployment of the facilities and network of KII-Government would also operate KII-Public. Investment and implementation of KII-Public was left to the private sector, with KT and Dacom carrying out the implementation as private operators.

The KII initiative, carried out between 1995 and 2005, built a network infrastructure that provided a diverse and extensive set of public services to the people and public institutions of Korea.

With the rapid increase in online activities such as e-commerce, Internet banking, and citizen engagement in public policy, Korea was poised to make a swift transition to the digital economy. In the e-government area, the former national digital network was completely replaced by the KII, allowing for the implementation of ambitious projects that had hitherto been impossible to execute. An E-Government Roadmap was established in August 2003, with 31 tasks to implement. The roadmap built on and followed the 11 major e-government projects selected in 2002. To deliver the public services set out in these 31 tasks, robust broadband service was essential. In fact, Korea possessed the necessary environment to provide such e-government services faster than other countries. It should be understood, however, that it was the Korean people who contributed the most to making use of such services. The government implemented a program for Bridging the Digital Divide; this included a project to provide IT Education for 10 Million Citizens[7] to not only address the digital-divide problem but also foster an environment that would promote the use of digital services through the Internet by all Koreans.

In conclusion, the KII was launched as a result of collaboration among the industry, citizens, the market, and the government, while investment and policy gave the impetus for e-governance-based reforms.

Software Infrastructure

Korea ran into a number of problems during the implementation of its e-government systems. First, the development of e-government systems based on open source, as adopted by the government, resulted in much difficulty during the initial stage. It is true that the development of software based on open source means being independent from various vendors, with fast and effective implementation of results gained from Business Process Reengineering (BPR) and Information Systems Planning (ISP). The development of such systems by different government agencies, however, led to problems concerning interoperability during the integration process of the differing systems. In addition, software provided by a particular vendor may also have interoperability issues, while the free development environment, under which open source-based software is developed, entails high costs when trying to meld such software, processes, and/or systems together.

During the nascent phase of e-government, a compatibility standard for data, security, and network was developed and enforced for e-government projects, but this measure did not even come close to guaranteeing compatibility between the massive number of systems operated by various government agencies. The government then implemented a number of tasks to guarantee compatibility between systems, especially the software being implemented and developed by government agencies, to put an end to the redundant development of various systems and to make it easier to improve or change systems.

By including the Information Technology Architecture (ITA), which was implemented from 2004 in the 31 e-government tasks, the government found a way to increase the effectiveness of investments through efficient management of IT resources.

The ITA policy was a type of reference model that analyzed the various elements such as business, application, data, technology, and security that make up a large organization, and structurally organized the relationship between these elements. A tool to support this policy, called the ITAMS (information technology architecture management system), was developed and distributed to various government agencies and public institutions. In 2005, the relevant laws were revised to make the implementation of ITA in public institutions mandatory. In fact, the government applied the ITA to the Government Integrated Data Center (GIDC) in 2005, making it the first case of ITA implementation being applied to government infrastructure. By applying the ITA, the government made it possible to develop software efficiently, as the implemented system enables recognition of redundancy and reusability of application, data, hardware, and software resources.

If the ITA policy can be called a reference model for IT assets, the e-Government Standard Framework can be called the infrastructure of components for developing software. In 2007, the Korean government carried out ISP to develop a common platform for software development based on open source. From November 2008 to 2009, the Korean government invited 10 companies to participate in the development process and announced the completed platform as the e-Government Standard Framework, which was based on Java.[8] The e-Government Standard Framework became the general structure based on open source, which eliminated dependency on a specific vendor by using public technology. It also presented a possible standard for connecting different commercial software and made systems replacement easier due to the modular structure of each component. As of 2014, some 450 e-government projects had made use of this framework, and more than 4,700 developers have been trained with more than 350,000 counts of download. The framework is fast becoming the infrastructure used to develop software in the public sector. It has been used in many different e-government projects in different sectors such as public data sharing projects, trade, and logistics projects, health insurance portal system development projects, and others, amounting to approximately US$1.263 billion in size.

The government also led the implementation of the Public Key Infrastructure (PKI) as another software infrastructure component. To support the safe use of e-government services, the implementation of a PKI became a policy that the government fully supported. A pilot operation center was established within the NIA in 2001, and the Government Public Key Infrastructure (GPKI) for e-government services was implemented based on open source and standards set by the Internet Engineering Task Force (IETF). All civil servants were given certificates for an electronic signature to enable authentication and authorization to log on to government systems. Currently, most of the e-government services provided by the government, including public services, make use of certificates for electronic signatures and pin codes for encryption to control access. A tele-working service is also being offered through the implemented GPKI.

Data Infrastructure

The data infrastructure of the e-Government of Korea was developed in several stages. When the NBIS was first being implemented in the 1980s, databases that could be used by computers were constructed by digitalizing paper documents for five public services such as resident and automobile registration. At the same time, a computer (TICOM) capable of using such databases was developed and distributed to public institutions, and data centers were established for government agencies. A separate government network to facilitate communication within the government was built, but due to limits in bandwidth and technology, there was difficulty in using this network for online public service delivery or internal government work. The government realized that possessing more high-speed bandwidth was a more pressing and important need than developing e-government services and began to make a large-scale investment in the high-speed KII-P for public services, including e-government services. Once this network infrastructure was set up, the ITA policy was implemented to guarantee compatibility and to integrate the services and systems of the government. The e-government framework was distributed to lower the cost of software development and eliminate vendor dependency for public projects. However, the core of compatibility and integration is the data.

Government Integrated Data Center. One of the foundational measures undertaken by the Korean government for building data infrastructure was the establishment of the GIDC. With the GIDC, all the data centers operated individually by different government agencies were integrated. During the integration process, the ITA policy was applied, government processes were integrated, and infrastructure was built to manage data that was previously separately managed by individual government agencies. This action called for placing an infrastructure that could produce, store, protect, manage, and distribute data from one physical location. The construction of the GIDC began in 2002, and it opened in 2005. It has become the core infrastructure for efficient use and management of data, with more than 20,000 pieces of hardware equipment installed, 1,200 types of work processed, 3.7 seconds of monthly downtime, a 30 percent reduction in data center operation cost, and the sharing of G-cloud[9] and Big Data platforms. By providing integrated support for common administrative work services, the efficiency of government agencies has increased significantly. Moreover, it provides the core functions of disclosing and supporting the use of public data possessed by the government by operating the government public data portal for open data.

Electronic Document Distribution System. Another achievement related to the implementation of data infrastructure is the implementation of the electronic document distribution system. To link the systems of the central government agencies and local governments and to expand the capacity for handling large amounts of documents, a project to improve the electronic document distribution system was launched in 2002. The system digitally produces, registers, distributes, manages, and stores government paper documents. The Electronic Document Distribution Center for standardizing the electronic document system and relaying the electronic documents changed the method of document distribution from message queueing, which pulls the document from storage, to the ebMS[10] method, which pushes the document out based on predefined protocols assigned in the ebMS registry. With the implementation of this system, more than 850 public institutions have been using the Electronic Document Distribution Service since 2008, and in 2008 alone, 47.48 million counts of electronic documents were transmitted and circulated, making it an infrastructure that achieved nearly 100 percent online distribution of government documents.

National Archive System. The National Archive system is the infrastructure that allows the use of data by automating and standardizing the whole process of gathering, preserving, utilizing, disposing of, and sharing information and records of public institutions. Specifically, the Business Reference Model (BRM), which began as a project for simplifying the redundant and complicated government work processes, and the record classification system were integrated to allow digital management of records produced by government work. The functions of record management and document preservation made it possible to connect the new record management system to existing electronic documents. The ability to extract the metadata needed for an integrated search of national records led to the implementation of the Meta Data Registry (MDR) in 2008. Extensive training for 771 public institutions took place in parallel, to deepen the understanding of various records, based on the new work classification system implemented after the inception of this system. The government also developed an open data platform in 2012 and released reams of public data to the general public through application platform interfaces (APIs) and standardization of data by sector.

Governance Infrastructure

For the Korean government, the most difficult issue during the implementation of e-government was the construction of the governance infrastructure. Implementation of e-government projects is characterized by the necessities of collaboration between multiple government agencies and other relevant organizations (Kim and Jho 2005), as well as a guarantee of project continuity for a minimum of 10 years. Such projects are extremely difficult because project owners need to have the power to change and coordinate government work processes, requiring budget appropriation and planning; at the same time, a governance system to advise and evaluate this whole process is also essential.

During the early years of e-government implementation, the National Information Society Agency (NIA) was established under the office of the prime minister. The NIA itself had a limited number of experts and supported the Ministry of Government Administration. With the inauguration of the MIC in 1995, all informatization projects, including e-government projects, became MIC's responsibility. The change was partly to establish and secure the necessary funds from telecommunication companies to support aggressive investment and also partly to make use of MIC's expertise in the implementation and utilization of IT systems, which other government agencies did not have. However, MIC did not have the power to supervise or manage all the e-government projects of other government ministries and agencies at the same peer hierarchy.

To address the problem, the newly created Informatization Promotion Committee (IPC) and the President's Special Committee for e-Government, made up of experts from academia, the industry, and government officials, carried out a review of policy decision making and prioritization for e-government projects. These committees were empowered to report directly to the president, giving them the authority for coordination. To enhance the expertise of these committees, the NIA provided technical support by reviewing various project proposals made by government agencies and, when needed, they took on the role of supervising pre- and post-project management, supporting project management, and designing pilot projects.

Three government agencies were selected and defined as core organizations that could coordinate the relevant organizations for the implementation of e-government: (a) the Ministry of Government Administration and Home Affairs (MOGAHA), responsible for local governments and administration; (b) MIC, responsible for fund management and strategy building; and (c) the Ministry of Planning and Budget (MPB), responsible for budget appropriation and execution. Such governance infrastructure clearly has many advantages, especially in implementing integrated government-wide projects. Specifically, it played a pivotal role during the implementation of large-scale national projects that required action no single government ministry could carry out, by coordinating and controlling opposition from and non-cooperation between government agencies for projects such as the e-Government Roadmap tasks and the construction of the GIDC, where the data centers of individual government ministries and agencies needed to be shut down. Although the structure and composition of the committees under each elected president were slightly different, it can be surmised that the maintenance of a similar governance infrastructure was the contributing factor to 20 years of consistent implementation of the Korean government's e-government projects.

Legal Infrastructure

The revision of legislation relevant to e-government was largely carried out in five areas: (1) legislation for digital work processing, such as the creation of electronic civil petitions and documents; (2) legislation to expand civil engagement, such as information disclosure and electronic voting; (3) legislation for prioritized projects, such as local e-government ordinances and the public finance information system; (4) legislation for building the e-government infrastructure, such as the GIDC; and (5) legislation to guarantee safety and reliability, such as the protection of private information possessed by public institutions. From 2003, more than 1,200 laws, enforcement decrees, and rules have been revised. After several years of public hearings and legal reviews, in 2008 the Personal

Information Protection Act was confirmed, enacted, and enforced to prevent the abuse and leak of private information.

Although not directly related to e-government, a number of comprehensive laws were passed to enforce the application of ITA policy for public projects of a certain size or larger. They include laws on open data, the Act on Promotion of the Use of Information, and the Digital Signature Act. These laws are part of the government's efforts to build a legal infrastructure that responds proactively to the fast-changing environment of e-government, and they have contributed much to the stable and safe use of e-government services by the people.

Toward the Digital Economy along with Digital Government and Public Services

After the completion of the KII and full operation and provision of e-government services, a new concern cropped up. The ICT industry became the central pillar of the Korean economy, led by large-scale investment from the government in the 1980s and the 1990s and resulting in the deployment of a fast telecommunications network and widespread e-government services. At the same time, however, with the birth of the smartphone, the coming of the mobile revolution, high-speed wireless communications that rivaled the speed of wired networks, and the convergence of the PC and HDTV, the digital age that the government had to prepare for was moving on to a digitalized economy from the current information sharing and processing environment.

The leap in ICT brought about another industrial revolution, creating new services and new industries through convergence in every industry. For an export-oriented economy like Korea's, which has no natural resources, it became imperative to prepare for the radical changes on the horizon. Accordingly, Korea decided to convert the high-speed broadband infrastructure, the KII that had already been launched, into the BcN infrastructure that would integrate broadcasting/video, the Internet, and audio streams. At the same time, a plan was established to implement the USN (Ubiquitous Sensor Network) infrastructure that could create diversified services by linking with wireless networks such as RFIDs (radio frequency identification devices) and sensing technology. In 2004, an ambitious plan was made to deploy 100 Mbps of broadband service to 20 million households by 2010 (NIA 2014a). This strategy was based on the predication that broadcasting, audio, and Internet services, including e-government and related public services, would be delivered through high definition (HD) quality, which would require enormous bandwidth. Many countries believed Korea was investing too much into its broadband network infrastructure, but despite the outside concerns, Korea went one step further with the government-led Giga Internet Service Plan, in 2009. Giga Internet was introduced in 51 cities through the development of relevant technologies, network convergence, and implementation of the Intelligent Operation Network. In fact, a new plan will deploy 1 Gbps Internet service to all 87 cities in Korea by 2017 (NIA 2014b).

It should be noted that, as time goes by, it is the leaders in market competition, such as the telecommunications operators, content providers, and broadcasting companies, who tend to make the investment in the telecommunications network infrastructure, while the government is establishing strategies, supporting the development of new technologies, and implementing pilot projects. The government's participation allows the private companies to compete for dominance in the market through large-scale investment. Based upon the fast soon-to-be available information infrastructure, e-government services are being converted into a full-fledged one-stop service. Since the establishment of the GIDC, government work processes have been integrated, and databases and services are being integrated into the Public Cloud as part of the conversion process. All e-government projects are now implemented based on the Enterprise Architecture, raising the need to consider compatibility and connectivity for implementation.

The government has greatly increased the number of mobile services delivered to the people, as the number of Koreans making use of e-government services via mobile devices has started to grow on a massive scale. Online payment and work portal services were all developed to be provided on a mobile platform, and a verification system for a safe applications list was put in place. The government has developed and distributed more than 1,200 free mobile apps to the people since 2011 and has been promoting the use of SNS (social networking service) by public institutions. Specifically, a mobile security infrastructure that can also be used by the government was built by developing the necessary security functions to assure the delivery of safe mobile e-government services. The mobile e-government test center is also being operated so that the government can expand the delivery of the e-government services and make use of the vast information infrastructure.

Korea is strategically approaching its entry into the digital economy where an immense amount of data needs to be stored, analyzed, and converted into new values from the emergence of new technology such as IoT (Internet of things), cloud services, and Big Data. In 2013, plans calling for the convergence of IT with other industries were announced, and strategies, such as developing convergence technology and business models, for full entry into the digital economy were established. The plans included creating new markets through the convergence of IT with five major industries (automobile, shipbuilding and offshore plants, textiles, national defense and aeronautics, and energy) and five major services (food, education, health care, disaster and public safety, and transportation). This approach by the government is expected to create 230,000 new jobs and increase production output by US$46 billion by 2017 (NIA 2013). Korea is expected to produce US$1.4 trillion in GDP and export US$600 billion in 2015. Including semiconductors, ICT-related industries are expected to produce US$150 billion. In addition, the convergence of ICT and existing industries will bring about a 3–4 percent increase in overall national production and a 2–3 percent increase in the country's employment rate. Although the increases may seem marginal, the effects from the conversion to the digital economy and the cultivation of a knowledge workforce have much potential.

Since 2012, Korea has initiated a new agenda called Gov. 3.0, which integrates all public services, gathers all public data, and helps public service providers make better decisions. Beneficiaries can get much smarter services because Gov. 3.0 provides customized services for each individual through analysis of opened public data. In other words, so-called "open data" becomes a fundamental concept to support all these new paradigms. To make it happen, key infrastructures are a big data platform, clouds, and data centers. Korea is planning to build the third consolidated data center to enhance the current capabilities of handling government clouds, which connect hundreds of government entities. The big data platform will be enhanced with more accessibility and availability of public data through cloud infrastructure. Mobile government services will connect customers and customized services to provide faster accessibility. It is obvious that the infrastructure for e-Government is evolving around "data" and is contributing to the digital economy.

Resolving Challenges and Conflicts

There have been many difficulties and challenges during the last 30 years of implementing the e-government infrastructure. Some of the problems in the nascent stage of the 1980s included workers' inexperience with using computers, the lack of an information infrastructure such as data networks, and limited time and budget to digitize the vast collection of paper documents. Another issue was the lack of relevant laws, dedicated departments, and human resources to implement e-government projects. It was the 1990s when the Korean government realized that there was no information infrastructure sufficient to start full-fledged delivery of e-government services. This became the period of building the much-needed information infrastructure.

Perhaps the most difficult issue was securing the budget and funds sufficient to realize such a vast endeavor without foreign investment and depending only on the government's budget. The government addressed the problem by establishing a fund based on the revenues of the telecommunications companies, but it had to persuade the companies to cooperate and gain consensus for using and managing the fund. To win the trust and support of the public for such a huge investment in the information infrastructure, the government started the IT Education for 10 Million Citizens initiative under the program for Bridging the Digital Divide, and also conducted a public campaign that included public promotions such as the Information and Culture Day.

Coming into the 2000s, during the full improvement phase of the e-government systems, the government still faced difficult situations. These included the following:

- *Finance system and accounting practices.* The Board of Audit and Inspection and other relevant government agencies began to oppose the introduction of a single public finance information system in accordance with international accounting standards. Because the introduction of an integrated finance system enhances control of the national financing process and achieves efficiencies, each agency realized it would not be able to keep its own finance system. Moreover, introducing single entry accounting in the new system was completely different from the old system and changed the accounting practices of each agency.

- *GIDC project and data issues.* All the government agencies that had to shut down their data centers were strongly against the project. A major complaint was that unnecessary work was being forced on government agencies in the process of applying the Enterprise Architecture to IT projects of government agencies, instead of the previous practice of applying such compliance rules only to large-size government projects. When tasks for sharing administrative information to provide better services to the people were being implemented, each of the relevant government agencies used various regulations and legislation to restrict access to their databases so that services requiring information from multiple government ministries and agencies, such as issuing passports, could not be provided as an e-government service.
- *Budget.* Budget appropriations caused friction between different government agencies during the prioritization process for major e-government services.
- *Human resource capability.* Many civil servants did not possess the capability to work under the new public service delivery system.
- *Procurement practices.* Redundant investment and problems in efficiency cropped up because of competitive procurement practices between different government agencies.

The starting point of the solution to these problems and challenges was the governance infrastructure. The committee responsible for implementation of e-government impressed upon the president, the prime minister, and other national leaders the importance of e-government services as the core element for strengthening national and business competitiveness and supporting the economic lives of the people through continuous communication. This presidential committee analyzed the issues between different government agencies and used expert groups and specialized organizations to find and review alternatives that would resolve the problems. Results were achieved by retraining and directly involving the civil servants that had been reluctant to face the changing environment in the e-government projects.

All e-government projects were required to produce annual results, thereby identifying long-running and unproductive projects. The indirect effect of reducing budgetary support for nonperforming or underperforming projects and government agencies was achieved by making public the results of the evaluations of government agencies and their projects. Diverse methods, such as direct negotiation, feasibility analysis by specialized organizations, analysis of government work processes, supervision of the IT system, and a satisfaction survey of the people who received public services, were used to resolve differences between government agencies. Interventions by the Special Committee for e-Government and evaluations by the budget authorities were also effective in reducing the differences and led to collaboration between government agencies.

The most critical success factor was maintaining urgency and consistency for the e-government projects being implemented, so that all the tasks of every government agency could be implemented only under the framework and direction of a national informatization strategy developed for the entire country.

Lessons Learned from Successes

The investment and implementation of the Korean e-government infrastructure achieved success for many reasons.

1. The project had continuous full support from all the presidential administrations, despite changes in government after elections.
2. A well-established governance infrastructure made e-government projects efficient and resolved the conflicts among stakeholders.
3. With the clear expression and intention of making the ICT industry the core of Korea's national economic strategy, the government saw e-government services as an essential tool to grow the economy.
4. Investment in information, data, software, and other infrastructure for e-government was made solely with government funding, as was their operation. The Informatization Promotion Fund was built up from profits made by telecommunications operators instead of foreign capital. Preparations for development were made early on, starting in the 1980s.
5. Implementing public projects and developing the ICT industry at the same time created a virtuous cycle.
6. The development of open source-based frameworks reduced dependency on foreign companies.
7. Efforts were put into developing e-government services to make use of the information infrastructure already deployed.
8. The creation of a platform for government service integration was done by eliminating all government data centers and integrating them into the GIDC.
9. Ten million Koreans were educated in information technology through the government-led informatization education/training program from the late 1990s to early 2000s.
10. The full utilization of the high-speed broadband infrastructure is the result of the prioritization of e-government projects.
11. Specialized public institutions, such as the NIA, had the expertise, knowledge, and experience to make efficient implementation of short-term projects possible.
12. The government responded to a changing environment by establishing strategic plans every few years.
13. Prioritizing the major projects eased the burden of budgets and time-consuming efforts.

Bringing Government into the 21st Century • http://dx.doi.org/10.1596/978-1-4648-0881-4

Failures of the Korean Digital Government Policies and Implementation

Setbacks along the way are to be expected.

1. The government concentrated too much on domestic technology development and standards, such as wireless broadband Internet technology and midsize computer technology, as the core infrastructure instead of identifying and capitalizing on global market trends.
2. Redundant and over-investment made in the e-government infrastructure by each ministry was caused by the excessive competition to expand the business and civil service boundary of the organization.
3. Many pilot projects were abandoned, thus wasting resources, because of hasty predictions of future technology that the market was unprepared to supply.
4. The e-government project lacked detailed and careful post-evaluations that are necessary to prevent repeating the same mistakes as well as making redundant investments and/or streamlining underperforming projects.
5. Project duration was fixed at one year, so any delayed and/or unfinished projects were not allowed to secure the next phase of the project budget, which degraded the quality of the delivered services.

Notes

1. TDX is the name for the Korean Electronic Switching Device.
2. TICOM is the name assigned by the government to the first midsize computer developed in Korea.
3. The first comprehensive data network built by the government for early e-government service.
4. ETRI was founded in 1978.
5. Total amount invested by the public and private sector for broadband infrastructure for the public, private and research networks.
6. KII is the name for the Korean broadband infrastructure.
7. The program provided free training programs for digital have-nots, including seniors, prison inmates, low-income families, and people in rural areas.
8. JAVA is a computer programming language that is intended to run any platform.
9. G-cloud is a government cloud built by Korea to deliver a cost-effective national cloud computing platform, using the Platform-as-a-Service, (PaaS) model.
10. Electronic Business using Extensible Mark-up Language is the standards sponsored by OASIS (Organization for Advancement of Structured Information Standards).

Bibliography

Kim, J. 2000. *Revolution of Korea's ICT*. Korea: Nanam Publishing Co.

Kim, P. S., and W. Jho, eds. 2005. *Building e-Governance: Challenges and Opportunities for Democracy, Administration, and Law*. Korea: National Institute of Administrative Sciences, National Computerization Society.

Larson, J. F., and H.-S. Park. 1993. *Global Television and the Politics of the Seoul Olympics*. Boulder, CO: Westview Press.

NIA (National Information Society Agency). 1996–2014. "Informatization White Paper."

———. 2005a. "Korea Information Infrastructure." Korean Publication.

———. 2005b. "Past and Present of Korea's IT Policy." Korean Publication.

———. 2006a. "Action Plan II for Broadband Convergence Network." Korean Publication.

———. 2006b. "20 Years History Book of NIA." Korean Publication.

———. 2013. "Korea: The World's Best Partner for Digital Prosperity."

———. 2014a. "National Informatization White Paper."

———. 2014b. 20th "Anniversary Report on National Informatization." Korean Publication.

NIA (National Information Society Agency), and Ministry of Public Administration and Security. 2008. "E-Government Project White Paper."

Oh, M., and J. F. Larson. 2011. *Digital Development in Korea*. New York: Routledge.

Yoon, J. 2006. "Chapter 5, Republic of Korea: The Journey to u-Korea—A Policy Perspective." In *ICT Best Practices in Denmark, Estonia, Finland, the Republic of Korea, Sweden and Switzerland*. Austria: Rundfunk-und-Telekom-Regulierungs-GmbH.

Yoon, J., and M. Chae. 2009. "Varying Criticality of Key Success Factors of National E-Strategy along the Status of Economic Development of Nations." *Government Information Quarterly* 26: 25–34.

CHAPTER 4

Evolution of Digital Government Systems in the Republic of Korea

Jungwoo Lee

Introduction

The initiation of government computerization in the Republic of Korea can be traced back to 1987, when the national basic information systems were initiated including administrative networks and critical databases. But efforts to build digital government beyond simple computerization, which evolved into the current state of e-government, started years later. The crucial year for digital government development was 2001, when the Presidential Special Committee for e-Government (SCeG) was established as a public–private partnership to set the agenda and monitor the progress of government digitalization in the age of the Internet. This is the year when the central government began to make a concerted effort to synchronize its systems development activities across governmental functions and levels. Since then, the digital government systems in Korea have been nurtured and have subsequently evolved. This chapter provides an account of the development of those systems, including the strategies that have nurtured the systems and their evolution into the current interconnected digital government of Korea. The systems are described and presented chronologically, and discussed in relation to different developmental phases.

In analyzing the evolution of e-government, a metaphorical story of developing information *islands* into information *archipelagos* and further into information *continents* is presented as a development framework of e-government. This metaphor is based on the natural tendency for information to be integrated and synthesized to make it more meaningful and usable. This metaphor of islands–archipelagos–continents (IAC) weaves the historical events of e-government development in Korea as it transitions from building functional information systems (islands), to connecting these functional systems to each other via network connections for information integration across different functions (archipelagos), and then forming a base platform of government information (continents) on which new and innovative citizen services are developed as necessary. The idea of evolution into information continents can be seen as an analytical lens to understand how digital government in Korea has been nurtured, so that it has grown into interconnected and integrated systems. This lens provides the basis for valuable insights into the future development of digital government, in which the nature of work both in and out of government will be changing at a fast pace.

Stages in Developing Digital Government Systems

The digitalization of government consists of large-scale programs and projects that require development by stages (Layne and Lee 2001; Lee 2010). The stages in digital government system development are larger in scale and scope than those found in the commercial sector, for several reasons: (1) the scope and scale of government operations are more expansive than those of almost any private organization; (2) the services provided by government cover every aspect of citizens' lives; (3) a high degree of complexity is involved in eliciting and implementing requirements from different levels of government as well as across functions of government, and in navigating through political processes among the various stakeholders involved in public decision making; and (4) digital government continues to evolve even after systems are put in place.

The history of digital government in Korea reveals five stages in developing digital systems for government (Song and Cho 2009; table 4.1). The 1st Stage (1987–95) started with the simple goal of computerization of public administrative processes. In terms of technology development, this stage coincides with the proliferation of personal computers, and the development of proprietary local and wide area networks and database technologies. In this context, the focus of this stage was on the development of nationwide secured but proprietary networks for administrative work and the construction of core databases containing information related to citizens, real estate, and automobiles.

Coinciding with the advent of the Internet and other network technologies, concerted efforts were made in the 2nd Stage (1996–2002) to enhance the completed network and further develop the databases that had been started in the 1st Stage. Nationwide high-speed network infrastructures were installed at this stage, and departmental and functional e-government systems were developed, albeit in a scattered manner across different levels and functions of government such as procurement, passports, patents and customs. In 1998, in the middle of the Asian Financial Crisis, Kim Dae-jung was elected president. During Korea's recovery from the crisis, his administration noticed the importance of e-government and established the SCeG under the direct supervision of the presidential office in 2001. The SCeG comprehensively reviewed ongoing information system projects across the levels and functions of government operations. After its comprehensive review, the SCeG launched 11 initiatives. These 11 initiatives were focused on the development and synchronization of digital government systems across functions and levels of government. This stage is termed the "full promotion" stage of e-government (see table 4.1).

Table 4.1 Stages of Digital Government

Developmental stages and metaphorical phases		
Stage	Phase	Administration
1st stage (1987–1995, Foundation): NBIS, administrative networks, digitization of national key databases including citizen registration and vehicle registration	Information Islands (Phase 1)	Chun Doo-hwan (1980–1988) Roh Tae-woo (1988–1993) Kim Young-sam (1993–1998)
2nd stage (1996–2002, Full promotion): Establish nationwide broadband networks; upgrade operational databases; launch SCeG in 2001; completed 11 major tasks defined by SCeG		Kim Dae-jung (1998–2003)
3rd stage (2003–2007, Diffusion and advance): Development of 31 key e-government projects including home tax service, e-procurement, Public Service 24 (G4C), and administrative information sharing system, etc.	Information Archipelagos (Phase 2)	Roh Moo-hyun (2003–2008)
4th stage (2008–2012, Integration): Integration and management of information systems of government agencies together; integration and linking of e-government services using cloud computing and hyper connected networks.	Information Continents (Phase 3)	Lee Myung-bak (2008–2013)
5th Phase (2013–2017, Maturity and co-producing): e-government 3.0; ICT innovation for service integrations; investment in IoT, Cloud Computing, Big Data for creative economy, ICT-enabled growth and jobs		Park Geun-hye (2013–2018)

Source: Adapted from box 1.1; Jeong 2006; Kim, Pan, and Pan 2007; Song and Cho 2009; and Lee 2011.

During the 3rd Stage of "diffusion and advance" of e-government systems (2003–2007), government-for-citizens (G4C) applications were developed and administrative information sharing systems were put into place. The 4th Stage (2008–2012), which focused on "integration," saw the launching of integrated e-government platforms. The 5th Stage of "maturity and co-producing" (2013–2017) is committed to innovation for service integration at all levels of government and investment in ICT-enabled growth through working with the private sector and engaging citizens.

This 5-stage model is the common depiction of the historical development of e-government in Korea. However, this chapter describes this evolutionary development of Korean e-government using the lens of the IAC framework in order to provide insights into the various technological stages. Table 4.1 presents the 5 stages, along with the metaphorical phases in the IAC framework and changes in the presidential administration in Korea.

As can be seen from table 4.1, this 5-stage model of e-government development is largely based on changes in the presidential leadership. Compared to the later 3 stages, the first 2 stages do not coincide exactly with changes in administration because technological development was not progressing as fast as in the later stages. In Stages 3–5, each administration came up with plans for e-government development when taking over the leadership of the country.

From the technological perspective, the developmental efforts of Korean e-government went through three phases, as presented in the middle column of table 4.1. (a) Phase 1: developing critical systems to build *information islands*, (b) Phase 2: building more systems and connecting the information islands to form *information archipelagos*, and (c) Phase 3: establishing platforms to form *information continents*. Despite variations in the development strategy for each of the 5 stages, the overall direction maintained a sense of continuity in terms of integration of information from islands to continents. Table 4.2 summarizes the strategies and actions in the formation of continents of information and information systems.

Phase 1: Although system development was initiated in the middle of the 1980s, the SCeG was only established in January 2001, under the direct guidance of President Kim Dae-jung. In May 2001, the SCeG devised and announced the first government-wide development plan for e-government as a national agenda for the new century. The focus of system developmental efforts during this phase was to construct and develop critical systems for government operations as well as for services to citizens. Eleven initiatives were announced—nine related to systems development and two associated with infrastructure readiness, as can be seen in figure 4.1.

Phase 2: Five years into the Roh Moo-hyun administration, previously developed systems were enhanced, upgraded, and interconnected to increase the efficiency of government operations and provide better service delivery to citizens. It appears that the SCeG began to realize that integration and

Table 4.2 Strategies and Actions in 3 Phases of IAC

Phase 1: critical systems initiation

Age of Information Islands (from Chun Doo-hwan to Kim Dae-jung administration)

Developed critical functional systems for government operations as well as citizen services

Initiated concerted and centralized efforts to control and monitor development projects

Established the President's Special Committee for e-Government (SCeG) as a public–private partnership

Characterized by 11 initiatives launched by the SCeG

Phase 2: more systems and interconnections

Age of Information Archipelago (from Roh Moo-hyun to Lee Myung-bak administration)

Continued building critical functional systems for operations and services to citizens

Continued with the SCeG

Recognized the need for integration and interconnection of systems

Characterized by 31 priorities based on the e-Government Roadmap

Phase 3: platforms for smart services

Age of Information Continents (from Lee Myung-bak administration until now)

Continued to build and interconnect systems

Evolved digital government platforms with common enterprise architecture

Allowed access to platforms and information by citizens and other systems

Transferred strategic control and monitoring from the SCeG to the Ministry of Public Affairs

Promoted smart government, ubiquitous technologies, and smart cities

Source: NIA 2010; MOGAHA (formerly called MOPAS) 2015.

Figure 4.1 Eleven Initiatives in Phase 1: *Information Islands*

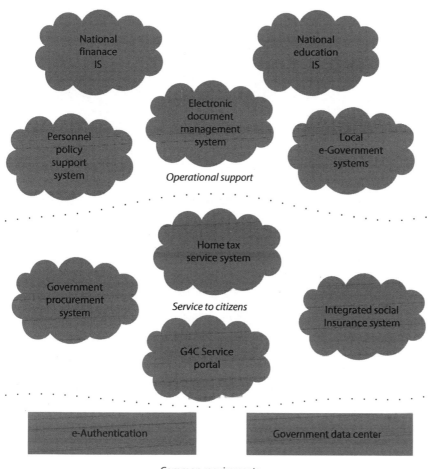

Source: Drawing based on NIA 2012.
Note: G4C = Government-for-Citizens, IS = Information System.

interconnection across systems would bring significant benefit to the public. Information, once stored in a system, has a tendency to seek and be connected to other information, creating synergistic effects. Thus, the information islands being built in Phase 1 looked for chances to get connected to other islands, forming information archipelagos. With the advent of the Internet, these connections were becoming technologically much easier than previously. In this context, the need for a government-wide common architecture was identified as important for further interconnection and linking systems. Thus, efforts were initiated to identify components of government-wide information technology architecture (ITA), later renamed and promoted as the GEA (Government Enterprise Architecture).

During Phase 2, 31 agenda items were identified, and investments were made accordingly. These included including upgrading current systems, developing new systems, and standardizing systems and services across different levels of government and functions (such as administrative information sharing, common operating reference models, and revision of laws and regulations). As a result, the idea of the government's ITA began to evolve with the goal of accommodating common reference models and a resource management framework.

Phase 3: Phase 3, which focused on "building information continents," actually began largely in the middle of President Lee Myung-bak's administration. As information islands are being connected to each other and information is being exchanged across islands, the overall picture of the Korean digital government becomes a collection of archipelagos. These archipelagos seemingly evolved into much larger interconnected systems, i.e. platforms integrating similar functions from different levels and branches of government. In this context, the focus of digital government shifted from technological development to government operations and services to citizens.

Catchphrases in this phase of information continents are mostly related to ubiquitous services and citizen-oriented digital government, signifying a service orientation. Thus, efforts were made to create new and value-added services, utilizing information available on the platforms of a variety of e-government systems. For example, a variety of ubiquitous services,[1] smart city services, and smart applications were developed and tested on these e-government platforms. The ITA initiated in Phase 2 evolved into the GEA, an e-government framework that standardized an open source framework applicable to all government-related systems and related libraries. In addition, modules were initiated under the title of the Government Enterprise Architecture Framework (GEAF).

In Phase 3, the focus moved from technological development toward service delivery for citizens and improved efficiency and ease of use for government officials, as some of the "continent" characteristics were practically realized and made accessible to the public and other systems. In 2011, the smart government plan was initiated with four strategic goals: openness, integration, collaboration, and green.[2]

The following sections provide more detailed descriptions of the systems, connections and policies that were developed in each phase—information islands, information archipelagos and information continents—to advance Korean e-government.

Phase 1: Developing Systems for Critical Government Functions—Information Islands

While getting Korea out of the turmoil created by the 1997 Asian Financial Crisis, Kim Dae-jung's administration recognized that the digitalization of government functions and citizen services was a key opportunity for government innovation. Consequently, the SCeG was established in January 2001, and the Electronic Government Act was enacted in March (Song 2002). In May

2001, the SCeG announced 11 initiatives for e-government development, to be completed in 2002 (SCeG 2003). During these two years, the focus was to construct and develop systems for critical government operations.

These 11 initiatives were selected based on careful reviews of ongoing computerization projects and the following assessment criteria:

- How closely is the system related to citizens' daily life?
- Does the system contain business process flows across functional boundaries in government?
- Does the database contained in the system have information sharing potential?
- Does the system streamline administrative processes? (Jeong 2006)

Overview of the Eleven Initiatives

The 11 initiatives in Phase 1 focused on developing critical e-government systems (figure 4.1). Specifically, 9 of the 11 initiatives were concerned with building functional systems, and the remaining 2 concentrated on common requirements across systems, namely e-authentication and the integrated government data center.

Out of the 9 functional systems, 5 dealt with efficient government internal operations, namely public personnel records management, public financial information, and government document management, as well as national level processing of education information and local government support. The other 4 were designed to enhance citizen services, namely the tax service, procurement and social insurance systems, and the integrated service portal for citizen services.

Most of these initiatives redefined existing computerization projects by expanding their scope and scale or integrating fragmented efforts with centralized planning and monitoring. For example, the integrated social insurance system initiative aimed to link the systems for pensions, health, accidents, and unemployment that already had been developed, implemented, and in use. Another example related to local e-government systems, which had been under development since 1998. These systems were fragmented and tended to focus on different functions. Thus, one integration initiative focused on the centralized selection and development of 21 government processes across 234 local governments. Last, the government procurement system was a critical integration project combining four systems developed separately: electronic document exchange (1997), online shopping mall (1998), electronic bidding system (2001), and electronic payment system (2001).

System Details in Phase 1

As already mentioned, among the 11 Phase 1 initiatives, 9 were systems, and the other 2 dealt with common requirements. Among the 9 systems, 5 were for operational support and 4 for enhancing citizen services.

Systems for Operational Support

Among a variety of government internal operations, five functions were identified as those that must be supported by integrated information systems. For these five systems, many small-scale system developments had already been undertaken by different functions or by different level governments. In general, these developments were scattered and fragmented and, thus, required integration.

National Education Information System (NEIS). NEIS is the system commonly used by all primary and secondary schools, educational district offices, and Departments of Education. NEIS connects more than 10,000 schools, providing parents and teachers with access to education-related information. The NEIS initiative resulted in the integration and upgrading of the student information system (which had been developed in 1995) and the school information management systems (developed in 1997). Stand-alone client server systems, which had been developed separately for each school, were integrated into a common database and networked interfaces using Internet capabilities.

National Finance Information System (NAFIS). Development of a national financial information system commenced in 1997 to correct the lack of common standards for collecting financial information and the need for real-time monitoring of financial activities across governments. The NAFIS was based on single-entry bookkeeping on a cash basis. The system connected accounting offices to the networked financial information of the Ministry of Finance and Economy for automatic summation and settlement processes. The Phase 1 initiative included the upgrading and integration of accounting and financial information systems in line with double entry bookkeeping and accrual basis accounting, which would allow for a common interface for information sharing among different agencies.

This initiative evolved into "dBrain" (Ministry of Planning and Finance 2015), a system that handles not only financial matters but also program management, revenue management, budget management, fund management, national property management, performance management, accounting settlement, and statistical analysis. The system connects 23 finance-related subsystems to provide real-time financial information for integrated budget planning, allocation, accounting, and settlement.

Personnel Policy Support System (PPSS). Earlier human resources systems were primarily paper-based. These systems were widely known to waste resources, have incompatible data, allow for inefficient maintenance, entail costly maintenance fees and experience difficulties in aggregating data. From November 2000 to August 2001, the Civil Service Commission developed the PPSS to improve human resource management, wage management, recruitment, education and training, statistics, and internal services for civil servants. With this new initiative, all personnel records of government officials have been integrated. The new system is now called e-People, and it is interconnected with the human resource management system and business process management system.

Electronic Document Management Systems (EDMS). The electronic document (lifecycle) management initiative was aimed at establishing a secure mechanism to electronically manage the entire flow of government document handling including production, approval, delivery, and archiving. This initiative integrated two different systems developed by two different agencies, the Ministry of Public Affairs and Security (MOPAS) and the National Archives of Korea (NAK). MOPAS had developed an electronic document processing and approval system in 1998, standardizing electronic document processing, and NAK had developed records management and archival systems.

Under this new initiative, the Government E-Document Distribution Center was established in September 2003. Distribution systems and archival systems were housed in the center, and linkages were made to all other systems, such as the business management system, the government function linkage system, and the digital budget and accounting system. As of August 2009, 1,283 agencies were using the EDMS, including 45 central government offices, 248 local administrations, 294 education offices and national and public universities, 342 public institutions, the Constitutional Court, and the Central Election Management Commission.

Local e-Government Information Systems (SAEOL). Since the local government elections in 1995, which revived the local government system, local governments with elected officials had been making efforts to develop information systems for their administrative use. In order to integrate these scattered efforts, an integrated local e-government information system, called SAEOL, was developed to handle administrative functions related to rural affairs, environment, and social welfare by connecting relevant agencies and systems. The new system was launched at the end of 2002. Following a number of upgrades and interconnections with other related systems, the modules now include land registry, farming, environment, complaints, health and welfare, local industries, residents, vehicles, finance and tax, construction, local development, culture and athletics, water and sewage, stock farming, forestry, fishery, roads and traffic, operational support, civil defense registry, family register, and disaster management. As of 2015, 26 of 31 functions identified in this initiative have been implemented.

Enhancing Services to Citizens
The following four systems were identified for development to enhance services to the public.

Home Tax Service (HTS). From 1999 to 2000 an electronic tax filing system was developed for the filing of personal income tax and value-added tax returns. This system was upgraded to include liquor tax, securities tax, stamp tax, and a special excise tax in 2001. The HTS integrated these systems, and now citizens can use the HTS system for online income and other tax administration services for businesses and individuals. Individuals can file tax return forms and request tax-related certificates at home.

Social Insurance Information System (SIIS). In 1989, the National Pension Corporation developed an information system to manage pension registration, collection, and benefits. The (un)employment insurance system and the industrial accidents insurance system were developed by the Ministry of Labor in 1995 and 1996, respectively and separately. The SIIS interconnected these insurance systems by linking pension, health, accident, and unemployment databases, and by providing a one-stop service. With this interconnected system, inquiries, civil petitions, notices, and payments of all four kinds of insurance can be processed online.

Korea Online E-Procurement System (KONEPS). In 1997, a document-exchange system with authentication was developed and implemented for the government procurement office, using electronic data interchange (EDI) technology to link corporations, public enterprises, and the government procurement office. Electronic bidding systems were established in 2000 and payment systems were added in 2001. The new Phase 1 initiative integrated these systems into KONEPS and opened up an integrated shopping mall-style two-way platform. Currently, with constant upgrades, KONEPS includes a customer relationship management (CRM) service, an intelligent product search service, electronic bidding by smart phone, and other functionalities.

Citizen Service Portal (Minwon24). In the mid-1990s, when the Internet and electronic commerce were in demand, ministries and agencies were pressed to offer online communication portals to the public. In 1997, the government's official website was launched, and online public administration services began, providing various forms and guidelines on services to citizens via the Internet. In 2000, the Home Citizen Service System was launched to enable citizens to send online requests from a personal computer and receive requested documents at home via e-mail. In 2001, the plan for citizen-oriented reform in public administration services was included as one of the 11 initiatives. The first services to be included in this phase of information islands were five major citizen services: residents, land, businesses, tax, and vehicles. These were incorporated into the Citizen Service Portal or 24/7 Citizen Service.

Providing for Common Requirements

E-authentication. E-authentication dealt with personal identification across different e-government systems, and the data center initiative integrated servers and applications from around different ministries and agencies in order to improve management efficiency and create opportunities for integration.

Government Data Center. The government data center initiative launched in this phase led to the successful construction of the integrated government data center in 2005 and the consolidation of hosting for the systems of 24 ministries in 2006.

Phase 1 of e-government largely came to an end around 2002. During this phase, some initiatives such as the government data center were initiated but not completed. Also, for political reasons, most of the systems that had been planned were proceeding with a degree of haste. However, despite the haste, the systems built during Phase 1 became a strong foundation for ongoing systems development.

Phase 2: More Systems and Interconnections—Forming Information Archipelagos

In 2003, the new Roh administration took office. Administration officials understood the importance of e-government and kept the SCeG in operation throughout the administration. Continuing the 11 initiatives initiated under the previous administration, the SCeG's first task in this phase was to produce a roadmap for e-government development. In May 2003, based on this road-map, 31 priorities for the next five years were proposed and announced by the SCeG.

E-Government Roadmap

The vision contained in the e-Government Roadmap in 2003 was to create the "world's best open e-government" (SCeG 2003). The roadmap outlined specific performance indicators to realize the vision. While the 11 initiatives in Phase 1 were defined around critical functional systems, the 31 priorities in Phase 2 were defined more broadly, setting goals and expanding the scope to encompass different systems across multiple government functions and levels. The 31 priorities are graphically presented in figure 4.2, along with the 11 initiatives in Phase 1.

The 31 priorities were designed to address the issues and challenges identified in executing the 11 initiatives in Phase 1. Once various e-government systems were established and operational, a big picture analysis was conducted and further requirements emerged. In particular, there was a need to interconnect and integrate these systems with other systems in order to fully leverage the value of information in each system. It was recognized that the value of information only increases when interconnected with other information.

Consequently, vertical and horizontal interconnection of systems across different levels and functions became a key issue in Phase 2. For example, systems such as the e-audit, e-assembly (an administration system for the National Assembly), and criminal justice systems were supposedly segregated from public administration in theory, as legislative and judiciary functions are supposed to be independent from administrative functions, so as to ensure a regulative balance between them. In the virtual world, however, these systems are connected to each other and share information for the convenience of information processing. The practical need of information sharing seems to be different from the theoretical principles of segregation and balancing. Databases are shared among these systems and are horizontally linked to systems on the administration side.

The consolidation of vertically separated functions was also planned for Phase 2 of information archipelagos. This involved the development of

Figure 4.2 Thirty-One Priorities of the e-Government Roadmap in Phase 2

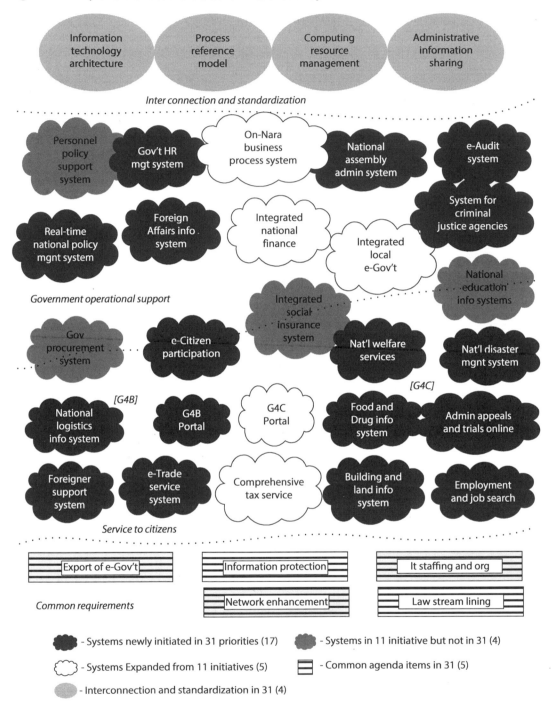

an integrated finance information system; the system included the local government finance information system as a subcomponent of a central government finance information system, which later became part of the digital brain project.

Another example of an information archipelago is the Personnel Policy Support System (PPSS) that had been developed as one of the 11 initiatives in Phase 1. In Phase 2, this became an important part of the government's human resource management system. These two systems also interconnected to the On-Nara business process system, providing detailed personnel information for communication and document processing.

System Details in Phase 2

The 31 priorities of Phase 2 included 22 information systems: 9 systems for internal operational support, and 13 for external services. Among these 22, 5 aimed to upgrade systems built in Phase 1 (part of the 11 initiatives) while 17 were new systems. Four initiatives related to interconnections across different systems, and five dealt with softer policy-related issues. The four interconnection-related priorities later evolved into the GEAF. One of the softer policy-related priorities involved the promotion and export of e-government solutions to other countries.

Because this chapter focuses on systems and related developmental efforts, the five priorities that deal with policy-related matters are not explained here. The 22 systems with 4 interconnection-related priorities are discussed below in more detail. They include:

- 9 systems that deal with operational support;
- 8 systems that relate to services for citizens (G4C);
- 1 system that supports citizen participation;
- 4 systems that support government to business (G4B); and
- 4 systems that deal with interconnections and standardization.

E-Audit System for the Board of Audit and Inspection

The Board of Audit and Inspection applied to the SCeG for systems development, and its proposal was approved as one of the 31 priorities. Although it was included in the roadmap in September 2003, actual development began only in September 2004, after eliciting and gathering business requirements. First to be developed were four primary modules: electronic audit management, which supported electronic processing of auditing procedures; data collection and analysis, which allowed for continuous collection and analysis of auditing data; audit knowledge management; and the portal. These four modules were completed in December 2005, and by March 2006 all of them had merged into the Integrated Government Audit System. This system now supports coordination of all government audit plans, including internal audits and the field audit management systems, which in turn support field audit activities including communications among field audit team members.

Administration System for the National Assembly

This initiative started with the submission and processing of computerized documents (such as proposed legislation, national budget review, and administrative affairs inspection) for legislative activities. As of 2006, this system was used by 514 organizations. Via the electronic document distribution system as part of e-Assembly, organizations can request and receive documents electronically. The number of requests and submissions increased from 774 in 2004 to 28,000 in 2006. The e-agenda module handles collaborative working on digitized legislative documents, and the e-voting module handles the voting procedure on the floor and automatic recording. The portal interface of the system discloses the official activities of members, and provides information about the schedule and agenda of the National Assembly, and the minutes of its meetings.

Integrated System for Criminal Justice Agencies

The integrated system for criminal justice agencies connects the police, prosecutors, and the Ministry of Justice with a common interconnection interface operated by the central operation center in the Ministry of Justice. This integrated system later evolved into the Korean Information System of Criminal Justice Services (KICS), accessed through the Criminal Justice portal. In addition, the court system is also interfaced through the operation center, so that the whole process of criminal justice from investigation through indictment, trial, judgment, and closing are all tracked electronically. Currently, 394 types of information are processed and accessed through this integrated system

- The police provide 74 types of information including opinion reports, transfer reports, and crime history;
- Prosecutors provide 160 types of information, such as arraignments, non-indictment decisions, and warrant claims;
- The courts provide 95 types of information including warrant issuance and dismissal, trial data, and judgments; and
- The Ministry of Justice provides 65 types of information, notably imprisonment and probation data.

The general public can electronically inquire about the progress of a case and about court records at any time through the criminal justice portal (www .kics.go.kr).

Foreign Affairs and Trade Information System

As one of the 31 priorities, the Ministry of Foreign Affairs and Trade carried out the e-diplomacy project. In 2005, a document exchange system was developed to enable the ministry, embassies, and consuls abroad to have access to nonclassified documents. In 2006, a separate system was developed for access and exchange related to classified documents. Subsequently, 68 different and disintegrated servers were integrated into information centers in the United States and Europe. Ultimately, foreign affairs processes were integrated into the e-diplomacy, e-consul, and e-passport systems.

Government Human Resource Management System

The PPSS, developed as one of the 11 initiatives in Phase 1, was expanded by interconnecting it with newly developed local government human resource management systems. In this regard, the Government Human Resource Management System is an example of an archipelago connecting adjacent islands of the PPSS and local government systems. Local human resource systems handle personnel affairs among and across local governments. These personnel matters include compensation, benefits, contract/work history, performance reviews, training documentation, certifications, and security clearance status. A standard management system that would be commonly applicable to all local governments was constructed in 2005, and interconnected with the PPSS developed for the central government. The PPSS had been used by 250,000 civil servants from 69 agencies as of June 2006, with the usage rate up to 92.6 percent. In Phase 2, new functions were introduced, including performance assessment, e-appointment of civil servants, and senior executive management services.

Real-Time System for National Policy Management

This priority consisted of two primary modules: government knowledge management and work management. The knowledge management module was developed based on groupware developed and used by the presidential office, the Blue House. The groupware included functionalities such as daily journal recording, document management, and task management. Later, this system was replicated for government-wide policy management and archiving. The government work management modules consisted of business process management, reference model management, customer relations management, and performance management.

Integrated National Disaster Management

The purpose of the National Disaster Management System (NDMS) was to support the management of disaster prevention, preparation, responsiveness, and rehabilitation. Disaster monitoring systems were deployed to 10 government agencies, including the Ministry of Agriculture and Forestry and the Ministry of Maritime Affairs and Fisheries. In 2006, a cross-governmental disaster management network system was established to strengthen the partnership among 71 related agencies. The national disaster information center can be accessed at www.safekorea.go.kr.

Consolidated Building and Land Information System

This priority focused on interconnecting two different but related systems and upgrading the services provided to citizens by integrating information provided by these two systems. The first was the architecture administration information system (AIS), which had been developed in 1998. The AIS electronically processed building construction information; it also provided documents such as building registers, electronic blueprints, and information on related policies and statistics, all on a real-time basis. The AIS was deployed in 2001, along with the Architectural Decision Support System (ADSS).

The second system was the real estate information management center established in 2003 by the Ministry of Public Administration and Home Affairs. The center monitored and managed land and building ownership; it also maintained a system through which real estate information of government agencies was collected and maintained.

A building register management system was established, connected to other related systems, and deployed to local governments in 2007. With this project completed, administrative work related to architecture approvals was able to be electronically processed through an architectural administration website, which had been established in 2004.

Integrated National Welfare Services

This priority was geared toward providing national welfare services online in an integrated manner, especially to socially vulnerable groups such as the handicapped, children, women, and the elderly. This Integrated National Welfare Information System consisted of several subsystems: the social security information system, the social service electronic voucher, the integrated childcare information system, the local health and medical information system, the social welfare facility information system, and the vulnerable social group support system. It has gradually expanded since 2003. The health and welfare portal (www.bokjiro.go.kr) began operation in 2005. In 2009, the Korea Health and Welfare Information Service was founded to manage the system and related services. In 2013, as part of the move toward integration, this Integrated National Welfare System was further interconnected with the welfare services modules of local e-government systems.

Integrated Food and Drug Information Service

This priority was set to develop and integrate two different but interrelated systems related to citizens' health and safety: the food and drug information system and the agricultural, livestock, and marine product safety information system. These information systems are now integrated into the portal of the Ministry of Food and Drug Safety, which provides information on food and nutrition; agricultural, livestock, and marine products; medical devices; medicine and drugs; biological material; and food safety. The following website has more information on the system: Refer to http://www.mfds.go.kr/eng/index.do.

Consolidated Information Service System for Employment and Job Search

This priority consisted of three projects: an integrated database for labor markets, a citizen portal for unemployment insurance and job search, and an analysis system for the national labor market. Primary networks for employment—including Work-net, an employment insurance network, and Hrd-net for job training—were interconnected, enabling citizens to search in a one-stop portal. To maintain and manage these systems and services, the Korea Employment Information Service (http://eng.keis.or.kr/eng/about/about_keis.jsp) was founded in 2006 with tasks such as occupational research, vocational counseling, overseas work placement, and integration and provision of employment information held by various employment service agencies.

Administrative Appeals and Trials Online

Since 1985, the Ministry of Government Legislation (MOGLEG) has been responsible for administrative rulings associated with violations of citizens' rights by government agencies. As the number of cases increased exponentially, it was determined that the paper-based process needed to be converted to an information system in Phase 2 of e-government. System development began in 2004 and was implemented in 2006, including the functions of registering and notifying, the ruling process, knowledge services, and so forth. This system evolved into an administrative trial hub portal (www.simpan.go.kr).

One-Stop Business Support System

The business support portal (G4B.go.kr) provides a wide range of information and services to support business activities such as civil service information, policy information, and additional services via a single online window. Detailed information on 1,887 corporate services and industrial information from 205 different organizations related to business operations, such as government approval, certification, and help on business matters, have been provided through this integrated portal since 2006. Various additional services essential for corporate activities are provided by linking to the national backbone networks, including the procurement, tax, and four major social insurance networks.

Nationally Integrated Logistics Information Service

This initiative aimed to improve logistics processes and services by standardizing logistics documents and increasing the information-sharing across government agencies. In the initial stage, a single window was established by using existing systems such as KLNet and KTNET, which enabled the sharing of documents submitted to the Ministry of Maritime Affairs and Fisheries, Korea Customs Service, Ministry of Justice, and National Quarantine Station. Additional functionalities such as harbor logistics, customs clearance, exit/entry control, quarantine control, and air logistics were gradually integrated, and evolved into the National Logistics Information Center, which now also hosts logistics decision support, multidimensional analysis, and administrative support, as well as a national portal.

E-Trade Service

Following the Trade Automation Act in 1991, the Ministry of Knowledge Economy made efforts to automate export-import operations. In 2003, the national e-trade committee was organized under the prime minister, as e-trade had been prioritized in the e-government roadmap. The committee worked on a proposal for an Internet-based open network to support the seamless processing of trade operations, such as marketing, foreign exchange, customs clearance, and logistics. This priority evolved into the current system of uTradeHub, which is composed of TradePortal, FTA Origin Management System, Logistics Portal, Banking Portal, and cTradeWorld (customer clearance system).

Foreigner Support Service

The Ministry of Commerce, Industry, and Energy, the Ministry of Labor, and Ministry of Justice worked together to develop a Foreigner Support Service (hikorea.go.kr). The three business areas (foreign investment, entry/departure/ stay, and employment) were covered in the initial plan, which was devised in 2005. This priority was designed to help foreigners living in Korea to extend their stay and to look for employment opportunities. Today, the system provides investment promotion and customization. Investment information, including industry and the local environment, is accessible through the hikorea portal. The online foreigner support service delivers documents, such as those related to employment for foreigners, and permits for entry and exit. It provides matching services between Korean companies and foreign investors. The system includes 41 types of information such as foreign investors, tips for living in Korea, immigration, and employment.

E-Participation by Citizens

This priority included two different projects: an online citizen participation portal and the disclosure of administrative information online. The citizen participation portal's goal was to integrate citizens' complaints, ideas, and policy participation into a one-stop single window service. This portal is now ePeople. It interconnects 47 central agencies, 144 embassies and consulates, 195 educational agencies, 244 municipalities, the court administration agency, and major public institutes. The second project, the government information disclosure project, was based on the Freedom of Information Act of 1996. Today, a one-stop portal (www.open.go.kr) for online administrative information disclosure enables citizens to request and receive information online without visiting offices.

Business Process System (BPS)

The EDMS, developed in Phase 1, evolved and expanded into a business process system called On-Nara BPS. On-Nara BPS handles administrative work processes in an electronic manner. Government business processes were standardized with a new classification system for administrative work. On-Nara BPS supports government officers by electronically processing their work, including planning, reporting, collaborating, approving, and executing. One of the features that may be having a stronger impact than originally planned is the memo report function, which is used to share information informally and to collect informal opinions. It is now being used by all government agencies at all levels.

Integrated National Finance Information System

In 1997, the government began the development of a national financial information system to address the lack of common standards for the collection of financial data and the need for real-time monitoring of financial activities across government. The system was completed in 1999. The system, which was based on single-entry bookkeeping on a cash basis, now allows accounting offices to be connected to networked financial information of the Ministry of Finance and Economy for automatic summation and settlement processes. An initiative in Phase 1 had included the upgrade and integration of current accounting and financial information systems in line with double entry bookkeeping and accrual basis accounting, allowing common interface for information sharing among different agencies.

In Phase 2, the single-entry bookkeeping-based financial management information system (FMIS) was upgraded into a double-entry bookkeeping system connected with other financial systems such as those involved with program and budget management. It was also interconnected with the financial management module developed in the local e-government system. At this point, the dBrain (MOSF 2015) project was launched. Today, the dBrain system connects 23 finance-related subsystems to provide real time financial information for integrated budget planning, allocation, accounting, and settlement.

Enhanced Local e-Government

An integrated local e-government information system for the basic unit-level municipalities, which was implemented in Phase 1, was to be expanded under Phase 2. The initial 21 targets for upgrades and enhancements were expanded to 31 in this phase, involving 234 basic unit-level municipalities.[3] These systems are now called SAEOL in Korean. Upgrades in this phase included the interconnections with other systems such as government finance systems, human resource management systems, and tax systems. Also, in this phase, 16 middle-level wide-area local government systems were planned, developed, and implemented. Interconnected with the basic unit-level municipality systems, these middle-level systems handled 24 functions. The Korea Local Information Research and Development Institute (KLID) was established in 2003 to manage the local e-government systems along with local finance systems, tax systems, and human resource management systems, all of which were interconnected with the central government and wide-area municipalities.

G4C Portal

The G4C portal implemented in Phase 1 with five major citizen services (residents, land, businesses, tax, and vehicles) continued to be a priority for system upgrade and expansion in Phase 2. G4C is now called the Citizen Service Portal. Currently, more than 3,000 kinds of applications for citizen services, such as certification and registration, can be made online, and 83 of them can be printed at home. Information can be browsed for more than 5,000 different kinds of citizen services, such as military service records, automobile tax status, local tax status, and so forth. Some citizen service information related to the life events of citizens, such as birth, home and care purchase, marriage, death, are handled by e-mail or short text messages. Applications increased to 68 million in 2011 from 30 million in 2007.

Comprehensive Tax Service

The goal of the consolidated online tax system was to make use of the Home Tax Service (HTS) more convenient and to enable taxpayers to deal more easily with matters related to tax notices, tax references, tax payments, and records online without visiting tax offices. Designed in 2001 as one of the 11 key initiatives, the HTS has been offering these services since 2002. Since then, the online payment of taxes has become one of the most frequently used services. In June 2002, the rate of e-notices reached record levels: 96.9 percent for corporate taxes and 81.2 percent for income tax. The kinds of citizen certificates issued online have also increased to 33 services. In addition, the online issuances of the six most used services, which account for 97 percent of citizen tax issuances, involved 643,000 documents online, which was 67.3 percent of the total.

Expanded Administrative Information Sharing

The administrative information sharing system (pr.share.go.kr) makes it possible for public agencies and financial institutions, as well as government agencies, to process applications that require documentation issued by government, without citizens having to submit an actual physical document. When a citizen applies for a service that requires government certification, the service provider can check the certification, such as resident registration and other related government-kept information, online. For example, when a citizen applies for a driver's license renewal, the issuing agency can check with the national health insurance service for the last health checkup record to assess the physical fitness of the applicant. Currently, this information sharing site offers 145 kinds of administrative information online.

Business Reference Model

The Government Business Reference Model (BRM) is a framework that is systematically structured according to government services and business processes. It classifies government functions by their goals and performance rather than the ministries in charge of such functions. It identifies all government functions and maps each of them with its related organization, laws, budget, and information systems. It reclassifies the functions according to goals, parties concerned, and service types for round-the-clock management. Government business functions were surveyed in 2004, and 17,800 businesses were identified, including 190 large functions, 660 mid-size functions, and 3,600 small functions. These processes were documented in this reference model. During the improvement and upgrade in 2006, the BRM of local governments was identified and developed into the functional classification system of central and local government agencies.

Integrated Computing Resource Management

As information systems and resource use proliferated across functions and levels, the need to record and maintain information about usage was recognized and implemented. Usage (including identifying duplication, promoting standardization, and enhancing procurement volume) had significant implications for investment and cost savings. Later, this resource management scheme was integrated into the architecture portal.

Information Technology Architecture (ITA)

This priority aimed to make all government agencies use a standardized ITA, supporting the systematic and standardized management of information resources and systems. The first pilot systems were implemented in 2005. Along with the pilot systems, a reference model was developed for smooth introduction of the ITA in terms of business processes and data. In addition, a standard ITA management system, which was developed in 2005, was offered to the agencies using the ITA. This priority led to the enactment of the Government Enterprise Architecture Act in December 2005.

Phase 2 of digital government was largely moved to Phase 3, the formation of continents, around 2010. The primary outcomes of Phase 2 were (1) extensive development of systems for almost every function and every level of government; and (2) the initiation of interconnection among systems and standardization of resources and interfaces.

Phase 3: Integrating Infrastructure and Interconnecting Systems—Formation of Continents

Phase 3 began in the middle of President Lee Myung-bak's administration. Lee's administration immediately restructured the organizational structure of central government. One notable restructuring concerned the Ministry of Information and Communication (MIC), which had been launched in 1994 by the Kim Young-sam administration to lead the strategic development of the ICT industry in Korea.

MIC's functions were broken up and split among the Ministry of Science and Technology, the Ministry of Industry and Resources, and the Ministry of Culture. Within this restructuring, the SCeG was demoted to the prime minister's committee from the presidential committee, and turned into an advisory organization without executive authority. The Ministry of Government Administration and Home Affairs (MOGAHA) took over responsibility for the planning and execution of e-government at that time.

The logical reasoning behind this restructuring was that information systems and technologies were mature enough to be converged with the actual functions; thus, the ministries themselves could take care of their IT-related development and promotion. Consequently, as most systems were developed, and interconnections and integrations were complete, the focus of digital government during Phase 3 also shifted from construction and development to utilization and services development. This was the reason why MOGAHA took over the planning functions of e-government systems from the SCeG.

In this regard, the National Informatization Master Plan, finalized and announced in November 2008, identified the goal of government digitalization as the achievement of efficient knowledge-based government, which provides new and integrated services to citizens and supports value creation for and by citizens and private businesses. It also aimed to integrate and interconnect the information systems of all departments and ministries to provide customer-oriented services while making government operations more efficient. In terms of quality, the plan aimed to increase the use of e-government systems from 41 percent in 2007 to 60 percent in 2012 and to move Korea up in the UN e-government index rankings to third place from sixth. In reality, the e- government system usage reached 60.2 percent in 2009, and Korea ranked first in the UN e-government index in April 2010.

The characteristics of Phase 3 can be summarized in the following three points, which are described in detail in the following section:

- from individual systems development towards standardization for the integration and connection of systems: GEAF
- from technological development towards services development: Smart Government Initiatives
- from building systems towards creative use of systems: proliferation of ubiquitous service development projects across municipalities and local governments.

Government Enterprise Architecture Framework

The interoperability of various digital government systems across functions and levels is not an easy task to achieve, and it has been recognized as a key challenge since 2001. The large-scale of government operations across different functionalities and services makes it difficult to standardize processes and procedures. Initially, each ministry built systems independently with specifications and solutions relevant to its particular needs. This approach did not give adequate attention to future needs to connect, exchange, and share data with other ministries and systems.

This silo effect resulted in a patchwork of heterogeneous systems with largely uncoordinated modules and components and interfaces. As systems upgraded for further transformation of government operations, the lack of interoperability appeared to be challenging, causing fragmented development of applications that would not be able to talk to each other, and challenging problems that went beyond simple hardware incompatibility.

In 2005, a law had been enacted to mandate the adoption of GEA, and this mandate not only covered post-hoc review, but rigorous reviews at the pre-planning stage. The review was to be conducted against common architectural criteria specified in the GEA. It took three years, from 2007 to 2009, to develop the detailed enterprise architecture (EA) standards, guidelines, and assessment tools for digital government into an integrated GEAF. The GEAF included the BRM, technology reference model (TRM), service reference model (SRM), and data reference model (DRM).

In 2008, the government EA portal (GEAP) was developed and became operational (www.geap.go.kr). According to the GEAP, 24,559 information systems were in operation as of December 2013: 8,216 for citizen services and 16,343 for government operations support.

As regards the GEAF, e-government framework components were also developed and distributed freely. Four types of eGovFrame are currently being provided via the e-government standard framework portal (www.egovframe.go.kr).

- The execution framework consists of five service layers that serve as common modules essential for the business execution environment: display processing, business processing, data processing, integration, and common framework.
- The operation framework provides communication tools for the efficient operation of information systems and monitoring tools for standard framework-based applications.
- The development framework provides tools for coding, debugging, testing, distributing, and configuring, all of which help the efficient development of e-government applications.
- The management framework provides systematic and efficient management functions, including service requests, modification, status, and standard management for application to IT projects (MOGAHA 2015).

Smart Government Initiatives

As most systems were interconnected and standardized via the GEAF and government-produced information was published, the creation of new values became the focus of digital government. In this regard, initiatives in Phase 3 were more diverse and dynamic. Open government was the primary philosophy. It was the catchphrase all along, and government at all levels initiated efforts to use and transform services creatively. Upgrading, enhancement, and interconnection of digital government systems in Korea became the norm rather than the exception. To provide integrated citizen-oriented services with synergistically interconnected information, the systems began to evolve on their own.

Through the Smart Government Initiative, which was announced in 2010 (MOPAS 2011), public users could enjoy easy and free access to government services regardless of the delivery channel. Use of electronic documents became a standard practice, and most administrative businesses, such as personnel management, finance, and procurement, were being handled electronically. All financial activities of the government could be managed in real time through the Digital Budget and Accounting System: dBrain (MOSF 2012). As of 2011, the KONEPS had some 44,000 public institutions and 220,000 suppliers as users, and became one of the world's largest e-marketplaces with a total transaction volume reaching 64 trillion won annually. The On-Nara business processing system is used by 362,000 workers in 103 government institutions, including the central ministries; and the time for business processing has been cut from 6 hours and 32 minutes to 3 hours and 27 minutes. The number of subscribers to the Citizen Service Portal increased from 3.59 million in 2008 to 9.21 million in 2011. The number of online certificate issuances also increased significantly to 28.24 million from 10.95 million during the same time period. The number of business information services provided through G4B increased to 3,941 in 2011 from 199 in 2005; and the average monthly number of visitors increased to 222,663 from 46,950 during the same time period (NIA 2013a, 2013b; MOGAHA 2015).

Proliferation of Ubiquitous Services Development Projects

Another interesting characteristic of Phase 3 is the development of ubiquitous services across different levels and functions of government. This development was feasible because most databases and systems were in place, with appropriate connectors standardized though the GEAF and eGovFramework (NIA 2013c). These databases and systems included:

- The U-Service Development Plan for Local Government, which was launched in 2007;
- U-Korea, which was set as one of 100 national agenda items in March 2008;
- IT Convergence, which was set as one of 17 new growth engine industries in January 2009;
- The e-Government Act, which was revised with ubiquitous based e-government development clauses in Article 18 in December 2009; and
- The Plan for Ubiquitous Technology-based Public Services, which was devised and executed by the National Information Society Agency (NIA) beginning in 2012.

Six areas were identified as critically related to e-government and therefore close to citizens' daily life: u-safety, u-life, u-infra, u-eco, u-admin, and u-tour. Instead of central control, these u-service projects were planned and conducted at the ministry and department level using databases and application interfaces open to the public. As of 2013, 228 ubiquitous services had been developed and tested as city-related services (Lee and Lee 2014). The Plan for Facilitating Ubiquitous Technology-based Public Services was established; it laid the legal foundation for adopting and utilizing ubiquitous technology-based e-government services after the revision of the Electronic Government Act in 2010.

In sum, Phase 3 of digital government in Korea started around 2010 and is still continuing in 2016. During this phase, initiatives were focused in general on service development and integration.

Implications: *Theory of Information Continents*

This chapter presents systems development in different but critical phases of digital government development in Korea from 2001 to 2012. The evolutionary model of digital government systems is presented along with the *theory of information continents, in which the focal transition of e-government developmental efforts is explained using the concepts of information islands, archipelagos, and continents.* This is presented as a metaphorical lens through which one can view concerted efforts to develop digital government in this era of Internet.

In the case of Korea, during Phase 1 of serious digital government development, most systems were developed as *information islands,* in which extensive databases were built and maintained. As time went by with these systems in operation, efforts were consolidated to interconnect these islands of information systems across different functions of government as well as across different levels of government, resulting in *information archipelagos.* As time passed with these systems in use, the archipelagos were interconnected with each other and formed tectonic plates of information—*information continents.* The progression from information islands to archipelagos to continents provides an analogy for the development of Korea's digital government system.

The historical events of digital government development in Korea, as presented in this chapter, illustrate the theory of information continents, where the connected systems in the final phase may work more like platforms on which new services can be designed. This theory is based on the strong synergistic orientation of information. When shared and linked to each other, the utility of information increases exponentially. In this regard, information systems have a tendency to be connected or integrated, or to converge towards each other, and digital government systems are no exception. This synergistic orientation is the reason that government processes can be reengineered in a way that could not be imagined before information technologies and the Internet.

After Continental Formation: *Continental Drift?*

It seems that once these information continents are formed, they make tectonic movement against other continents for the same reason that islands are forged into archipelagos and continents—synergistic orientation of information. For example, currently, the Real Time Cash Management System (RCMS) is under development for R&D programs in Korea (Han, Lee, and Lee 2014). The RCMS interconnects the project management system in government-funded R&D projects with the banking system. Via RCMS, funds are transmitted on an accrual basis in real time rather than through reimbursement or a prepaid scheme. Because of the crevice between the government continent and the banking continent, a few misappropriations in the cash remittance process have occurred. The RCMS is attaching the R&D information continent on the government side to the banking-related information continent in the private sector in real time. This tectonic attachment of these two information continents increases the transparency of government operations and, as a result, eliminates the misappropriations that had previously occurred.

Bringing Government into the 21st Century • http://dx.doi.org/10.1596/978-1-4648-0881-4

In the near future, we may expect integration and interconnections of the public information continent with the private information continent, as well as further integration within each continent. The public and private continents may not be distinguishable in the future, just as the citizen service continents are not completely separate from the internal operational continents of government. This theory of information continents has focused primarily on technological integration in terms of infrastructure and data, but it now needs to be expanded to include service and business components.

Conclusion

Developing the digital government system requires continuous and progressive efforts in order to nurture and grow systems and transform government processes and citizen services. The systems development process is not like constructing a building in which concrete is poured into shapes that supposedly last permanently. Information systems such as digital government need to be grown and nurtured like organic entities; they continuously evolve with the changing nature of work and of the services to be provided.

This idea of evolving information continents is proposed here as a powerful analytical lens to understand how digital government is nurtured and grown into interconnected and integrated systems and processes, ultimately leading to a versatile platform of information on which future services can be developed and current services transformed into new ones. In this chapter, Korea's experience in developing digital government is contextualized by this framework.

One thing to note is that this theory of information continents describes the past historical development of digital government. As the last stage of information continents is being proposed as a kind of open platform, the Korean digital government will evolve using these continental platforms. Extending the metaphorical analogy, the digital government of Korea from now on will experience some kind of movement of tectonic plates underground. It will be interesting to see how these tectonic movements impact the changing nature of work in public administration and governance of central and local governments in Korea.

Notes

1. The expression "ubiquitous services" was coined from "ubiquitous computing." As the dictionary definition of ubiquitous is "existing everywhere," it is often used to designate services that can be accessed or executed anywhere and everywhere, especially using information communication technologies.

2. Green ICT involves ICT with a low carbon footprint and that is environmentally sustainable.

3. Korean government consists of three levels: central government, middle-level wide-area local government, and basic unit-level municipalities.

Bibliography

Han, S.-Y., H. Lee, and J. Lee. 2014. "Developing a Real-time Cashflow Management System for National R&D Management." *Journal of the Korea Society of IT Services* 13 (3): 343–57.

Jeong, K.-H. 2006. *E-Government, the Road to Innovation: Principles and Experiences in Korea.* Gil-Job-E MEDIA.

Kim, H. J., G. Pan, and S. L. Pan. 2007. "Managing IT-Enabled Transformation in the Public Sector: A Case Study on e-Government in South Korea." *Government Information Quarterly* 24 (2): 338–52.

Layne, K., and J. Lee. 2001. "Developing Fully Functional e-Government: A Four Stage Model." *Government Information Quarterly* 18 (2): 122–36.

Lee, J. 2010. "10-Year Retrospect on Stage Models of e-Government: A Qualitative Meta-Synthesis." *Government Information Quarterly* 27 (3): 220–30.

Lee, J., and H. Lee. 2014. "Developing and Validating a Citizen-Centric Typology for Smart City Services." *Government Information Quarterly* 31 (Suppl 1): S93–S105.

Lee, Y. B. 2011. *2011 Modularization of Korea's Development Experience: The Introduction of e-Government in Korea.* Seoul: KDI School of Public Policy and Management, 138.

MOGAHA (Ministry of Government Administration and Home Affairs). 2015. *E-Government of Korea Best Practices.*

MOPAS (Ministry of Public Administration and Security). 2011. "Smart Digital Government Plan." Seoul, Republic of Korea.

MOSF (Ministry of Strategy and Finance). 2012. "DBAS, Korea's Integrated Financial Management Information System: Digital Budget and Accounting System." Seoul: Ministry of Strategy and Finance, 1–12.

———. 2015. "Digital Budget and Accounting System." https://www.digitalbrain.go.kr/kor/view/intro/intro02_01_03.jsp?code=DB04010103.

NIA (National Information Society Agency). 2010. "2010 National Informatization White Paper." (English).

———. 2012. *Electronic Government White Paper, 2008–2012* (Korean).

———. 2013a. "Korea Informatized Progress and Status Overview 2013."

———. 2013b. "National Informatization White Paper, 2013."

———. 2013c. "E-Government Standard Framework (e-GovFrame)." Seoul: National Information Society Agency.

Song, H. J. 2002. "Prospects and Limitations of the e-Government Initiative in Korea." *International Review of Public Administration* 7 (2): 45–53.

Song, H. J., and T. Cho. 2009. "E-Government of Korea: Achievements and Tasks." *Informatization Policy* 14 (4): 20–37.

SCeG (Special Committee for e-Government). 2003. *Korea's e-Government: Completion of e-Government Framework.*

CHAPTER 5

Digital Government Impacts in the Republic of Korea: Lessons and Recommendations for Developing Countries

Jooho Lee

Introduction

Since the 1970s, the Korean government has actively adopted information and communication technologies (ICT) as a strategic means of achieving the various goals of public policy and administration. In the 2000s, the government advanced e-government to provide citizen-oriented services and to engage citizens through two-way communication. And recently, it has initiated a new vision of e-government, called Government 3.0. The main strategic goals of Government 3.0 are to make the government more transparent, competent, and service-oriented by enhancing openness, collaboration, and two-way communication.

Scholars and practitioners in the global community have paid considerable attention to understanding the driving forces behind the success of the Republic of Korea's e-government development (Chung 2009). However, only a few studies have systematically analyzed the impacts of e-government in Korea. The aim of this chapter is to develop evidence-based e-government recommendations for developing countries, based on an assessment of the impacts of Korea's e-government initiative.

The chapter focuses on four dimensions of Korean e-government impacts: administrative, economic, political, and social.

- *The administrative impacts* of e-government are analyzed by examining the relationship between the adoption of e-government and organizational structure (e.g., downsizing), process (e.g., policy making), and outcomes (e.g., citizen satisfaction).

- *The economic impacts* relate to cost reduction (e.g., saving time, saving money, and improving performance) as well as the effect on both the national and local economy (e.g., ICT-related job creation).
- *The political impacts* are analyzed by looking at how e-government affects government responsiveness through electronic participation (e-participation) and accountability through enhanced transparency in government.
- *The social impacts* are understood by examining public trust in government, as well as social inclusion and cohesion.

The methodology for this study involves an examination of academic journals and documents and reports from government agencies and government-sponsored research institutions. The chapter classifies and assesses positive, limited, or unexpected impacts of e-government in the context of central and local Korean e-government. Taking a balanced approach to understanding e-government impacts facilitates the learning of appropriate lessons and the formulation of evidence-based recommendations for central and local governments in developing countries that are pursuing e-government initiatives.

Scope and Methodology

E-government is broadly defined as "a provision of government information and services using web-based technologies" (Moon 2002). Although some studies distinguish between ICT in government and e-government (Moon, Lee, and Roh 2014), we consider e-government to include both the inward applications of ICT in government for various internal operations and the outward applications of ICT for various services, including government-to-citizen (G2C), government-to-business (G2B), and government-to-government (G2G). Consequently, this study involves a review of Korean e-government studies dealing with both conventional ICT in government and e-government services.

A comprehensive literature review has been employed as the primary methodology. The process for selecting relevant literature involved three steps: (a) sources of articles on Korean e-government impact research were identified; (b) a deeper review was conducted to select articles that exclusively discussed the impacts of Korean e-government; and (c) a secondary review was conducted to ensure that the article was based on an empirical study.

As a result of this selection process, 23 empirical e-government impact studies were chosen for analysis. The findings of these studies were then classified into four broad dimensions: administrative, economic, political, and social impacts.

The Four Dimensions of Impacts

Administrative Impacts
This section considers the effects of Korean e-government on organizational structure and processes, as well as on organizational effectiveness and outcomes in terms of user satisfaction.

Organizational Structure

The relationship between ICT and organizational structure has long been discussed among scholars in the fields of management information systems (MIS) and e-government. To systematically understand structural changes, five dimensions of organizational structure were assessed: organizational power, centralization, formalization, red tape, and complexity (Rainey 2014). Tables 5.1 and 5.2 summarize the effects of Korean e-government on these dimensions of organizational structure.

- *Organizational power* is the decision-making authority within a hierarchical structure.
- *Centralization* refers to the extent to which the decision-making authority is concentrated at the organization's higher levels.
- *Formalization* is the degree to which an organization has formally written rules and regulations determining structure and procedures.
- *Red tape* is broadly defined as excessive and unnecessary administrative time delay.
- *Complexity* is the degree of vertical and horizontal differentiation and is measured by number of levels, sub-units, and specialization in an organization.

Some descriptive studies reported that the adoption of e-government applications has a positive relationship with centralization and formalization (Choi and Hahn 2008) and reduced red tape (Mok, Myeong, and Yun 2002). For example, Choi and Hahn surveyed employees of the National Tax Service to determine the impact of the Home Tax Service (HTS) systems. Forty-five percent of the respondents reported that the HTS systems centralized organizational structure, and 17.5 percent said that the HTS systems led to decentralization. Mok, Myeong, and Yun (2002) showed that 62 percent of survey respondents at seven state-level governments perceived the positive effects of ICT on reducing red tape in the government. Using survey data from 422 officials at central agencies, Yu, Kim, and Yoo (1994) found that (a) 55 percent of the respondents reported that computer use does not change the existing organizational structure, and (b) 74 percent perceived that computer use increased the power of the information technology (IT) unit in their organization.

The assessment found that some scholars in e-government had conducted multivariate analyses to examine the effects of e-government on sub-dimensions of complexity such as downsizing and span of control. As shown in table 5.2, one study considering downsizing found that the level of informatization had no significant impact on the size of local government, as measured by the number of full-time employees (Eom and Kim 2005). Other longitudinal research demonstrated that the adoption of ICT increased the size of 10 central agencies between 1989 and 2005 (Im 2011). Both studies found that ICT adoption at both local government and central agencies increased the span of control of middle managers, as measured by the ratio of middle managers to lower-level employees (Eom and Kim 2005) and the number of subordinates of middle managers (Im 2011).

Table 5.1 E-Government Effects on Organizational Structure: Descriptive Studies

Dimension	Effects	ICT applications	Positive response (%)	Negative response (%)	Research method	Reference
Centralization	Change in the existing power structure	General computer use	8.6	55	Survey of 422 officials at 10 central agencies	Yu, Kim, and Yoo (1994)
Organizational power	Increased power of ICT unit		74	5.3		
	Increased power of end-users	Home tax service	17.5	38.5	Surveys of 127 and 198 officials in 2003 and 2006, respectively	Choi and Hahn (2008)
	Reinforced power of top management		45	9.1		
Red tape	Reduced red tape	General ICT use	62	8.4	Survey of 405 officials at 7 state-level governments	Mok, Myeong, and Yun (2002)
Formalization	Increased formalization	Home tax service	64.6	8	Survey of 127 and 198 officials in 2003 and Survey of surveys in 2006, respectively	Choi and Hahn (2008)

Note: ICT = information and communication technology; IT = information technology.

Table 5.2　E-Government Effects on Organizational Structure: Multivariate Studies

Dimension	Effect	Independent variable	Finding	Research method	Reference
Complexity	Downsizing	Informatization index	Insignificant	232 local governments in 2002	Eom and Kim (2005)
		ICT use	Positive	A longitudinal study of 10 central government agencies between 1989 and 2005	Im (2011)
	Span of control	Informatization index	Positive	232 local governments in 2002	Eom and Kim (2005)
		ICT use	Positive	A longitudinal study of 10 central government agencies between 1989 and 2005	Im (2011)

Note: ICT = information and communication technology.

Organizational Processes

As to the effects of ICT on organizational processes, the literature mainly discusses four broad areas: decision making, coordination and communication, work processes, and knowledge quality and sharing. In fact, ICT has been touted as a useful means of improving decision making (by generating quality information and knowledge), enhancing coordination and communication within an organization, streamlining work processes, and increasing knowledge quality and sharing. Table 5.3 provides some descriptive studies and their findings in terms of the roles of ICT in these areas, while table 5.4 shows the findings of some multivariate analyses.

Decision making. An early descriptive study reported that many government officials (43 percent) do not agree that computer use in central government agencies automates decision-making processes (Yu, Kim, and Yoo 1994). A 2010 longitudinal study of ICT impact on decision making between 1998 and 2005, however, reported that local government officials in Korea's two largest cities, Seoul and Busan, perceived that ICT use improved decision making in terms of policy goal-setting (e.g., ICT facilitates the identification of related organizations' policies and plans) and searching for policy alternatives (e.g., ICT facilitates identification of social indicators and resources) (Myeong and Choi 2010).

As regards multivariate analyses, an early research project (Mok, Choi and Myeong 1998) found that local government officials improved two types of policy decision-making processes—policy goal-setting and searching for policy alternatives—through the use of ICT. However, the same study showed no support for the relationship between satisfaction with ICT service and improvement in decision making. Lim and Tang (2008) used a structural equation model to analyze a 2003 survey of 315 officials in 74 cities. They found that city officials made sound decisions on environmental policy when their cities provided high quality e-government websites, supportive IT leadership, and had access to citizens' online input on environmental issues. More recent research also provides evidence on the positive role of ICT applications on decision making.

Table 5.3 E-Government Effects on Organizational Processes: Descriptive Studies

Dimension	Effect	ICT applications	Positive response	Negative response	Research method	Reference
Decision making	Automate decision making	General computer use	26%	43%	1994 survey of 422 officials at 10 central agencies	Yu, Kim, and Yoo (1994)
	Policy goal-setting and searching policy alternatives	ICT use	7 of 8 indicators for goal-setting (e.g., identify related organizations' policies and plans); 5 of 6 indicators for searching for policy alternatives (e.g., identify social indicators and resources) show a significant improvement between 1998 and 2005		Surveys of 364 and 269 officials in Seoul and Busan in 1998 and 2005, respectively	Myeong and Choi (2010)
Coordination and communication	Coordinated service delivery	General computer use	27%	22%	1994 survey of 422 officials at 10 central agencies	Yu, Kim, and Yoo (1994)
	Enhanced communication		28%	25%		
	Improved interaction with clients		16%	42%		
Work processes	Faster work processes	General IT investment	48%	8%	2003 survey of 552 officials at 3 local governments	Han (2005)
	Simplified work processes		52%	9%		

Note: ICT = information and communication technology; IT = information technology.

Using 2008 survey data of 459 officials at 10 central agencies, Lim and Kang (2013) reported that officials' utilization of more functions and services provided by On-Nara systems is positively and significantly related to effective decision making.

Coordination and communication. Empirical studies fail to provide evidence supporting expectations of improved coordination and communication through ICT. Using 1994 survey data from 422 officials working at central agencies, an early study (Yu, Kim, and Yoo 1994) reported that only 27 percent of respondents believed computer use had positive effects on coordinated service delivery, while 22 percent considered such impacts to be negative. In a similar vein, the same study reported that only 28 percent of respondents perceived the impacts of computer use on enhanced communication to be positive. Moreover, this study reported that only 16 percent of respondents agree with the statement that computer use improves interaction with clients, while 42 percent of officials do not agree.

Table 5.4 E-government Effects on Organizational Processes: Multivariate Studies

Dimension	Effect	Positive impact	Insignificant impact	Research method	Reference
Decision making	Set up policy goal and choose policy alternatives	Perceived usefulness of IT as communication tool	Satisfaction with IT service	1988 survey of 513 officials at three state-level local governments	Mok, Choi, and Myeong (1998)
	Decision quality	E-government web-quality; online citizens' input; IT leadership		2003 content analysis of city government websites and a survey of 315 officials at 74 city governments	Lim and Tang (2008)
	Effective decision making	Diverse use of information systems; Leadership; organizational culture		2008 survey of 459 officials at 10 central agencies	Lim and Kang (2013)
Knowledge quality and sharing	Motivation of knowledge acquisition	Absorptive capacity; self-efficacy; perceived usefulness of knowledge; interpersonal trust	KMS quality; rewards; managerial support; culture of organizational learning	2003 survey of 176 staff members at a school district in the city of Gwangju	Kim (2004)
	Knowledge sharing capability	Use of IT applications (e.g., KMS)	Perceived ease of use	2003 survey of 322 employees at five public and five private organizations	Kim and Lee (2006)
	Quality knowledge generated by KMS	Leadership (e.g., interest in and support for KMS)		2008 survey of 2,275 employees at four public institutions	Lee (2010)
Communication	Communication with peer employees and parents and trust between teachers and parents	NEIS competency; information security and convenience	System quality	2007 survey of 1,440 teachers at schools in Korea	Song et al. (2008)
Work processes	Process improvement	NEIS competency; information security, convenience, and system quality			

Note: IT = information technology; KMS = knowledge management systems; NEIS = National Education Information System.

Work processes. Korean e-government studies appear to offer consistent findings regarding the effects of ICT on work processes. Han (2005) found that local government officials gave a positive assessment on ICT investment (e.g., ICT budget, ICT employees) in increasing the speed of work processes and in simplifying them.

Knowledge quality and sharing. As various ICT applications have been adopted to generate, store, and transfer data, information, and knowledge, e-government scholars in Korea have paid attention to the effects of knowledge management systems (KMS) on knowledge quality and sharing (Kim and Lee 2006). Korean e-government studies show mixed findings. Using a 2003 survey of 176 staff members at a school district in Gwangju, Kim (2004) found that perception of KMS quality (e.g., accessibility) had no significant relationship with the motivation to acquire the knowledge generated by the KMS. Kim and Lee (2006), however, reported that public sector employees' use of KMS had significant effects on increasing knowledge-sharing capability, although perceived ease of use had no significant relationship with knowledge-sharing capability. Lee (2010) examined the role of leadership in quality knowledge created by KMS. Using a 2008 survey of 2,275 employees at four public institutions (enterprises), this research found that respondents perceive high-quality knowledge when leaders in their organizations are interested in and support KMS.

Song et al. (2008) concluded that the perceived quality of the National Education Information System (NEIS) did not significantly impact enhanced communication with peer employees and parents. This study, however, found that teachers and staff at elementary and secondary schools perceived process improvement when they felt competent working in NEIS, when NEIS appropriately handled information security, when they perceived NEIS as convenient, and when they perceived that the system produced high quality results.

User Satisfaction and Organizational Effectiveness

End-users' satisfaction with information systems has received ongoing attention as one indicator of the success of information system adoption. In a similar vein, e-government researchers (e.g., Morgeson 2012) often use satisfaction with ICT applications as a proxy for measuring the organizational outcome of ICT adoption. This line of research focuses on two types of end-users—government officials and citizens/businesses. Another indicator of ICT adoption success is its contribution to improved organizational effectiveness. In general, this assessment reveals that end-users in government and citizen users are satisfied with ICT applications and e-government services.

As shown in table 5.5, Yu, Kim, and Yoo (1994) found that 89 percent of government officials at central agencies reported that they were satisfied with the service quality produced by computers in their organizations. Government employees reported greater satisfaction with ICT applications when they perceived that their organizations were equipped with a higher level of informatization (Lee and Oh 2000). In 2008, Choi and Hahn's longitudinal study of HTS applications to the national tax office revealed that government officials'

Table 5.5 E-Government Effects on Organizational Output: Descriptive Studies

Dimension	Effect	ICT applications	Positive response	Negative response	Research method	Reference
Quality of service	Perceived quality of service delivery	Computer use	89%	3%	1994 survey of 422 officials at 10 central agencies	Yu, Kim, and Yoo (1994)
User satisfaction	Satisfaction with ICT applications	Informatization	Positive		Survey of 328 officials at four state-level governments	Lee and Oh (2000)
	Satisfaction with Home Tax Service applications	Home Tax Service applications	22% (2003) to 57% (2006)		Surveys of 127 and 198 officials in 2003 and 2006, respectively	Choi and Hahn (2008)
Effectiveness	Perceived effectiveness of e-government services	Online parking applications	Strong cross-unit communication ties between IT, program units, and private IT vendor.		Survey of 174 officials at two district governments (Gangnam and Seocho) in 2005 and 2006, respectively	Lee (2013)

Note: ICT = information and communication technology.

satisfaction with HTS had increased to 57 percent in 2006 from 22 percent in 2003. Lee (2013) explores the effects of communication networks among IT and program units in a local government and IT vendors on their perception of the effectiveness of e-government services. Using social network analysis and employee survey data of 174 staff responsible for dealing with online parking applications and the IT units in two district governments (Gangnam and Seocho districts) in 2005, this study found that when local e-government services are provided through communication networks that involve a greater number of employees (i.e., the parking officers and IT unit staff) and when there is more frequent communication between the parties concerned (i.e., parking officers, IT unit staff and private IT vendors), local employees in those networks tend to positively assess e-government effectiveness.

Table 5.6 illustrates the findings of e-government effects on organizational output, based on multivariate analyses. Lee (2010) reported that public employees' perception of quality knowledge created by a KMS and system quality increased satisfaction with the KMS. Using a 2004 survey of 485 citizen users of local e-government services of the city of Daegu. Sung and Jang (2005) found that citizen users' satisfaction with local e-government services were significantly related to greater responsiveness, speediness, openness to communication, and reliability. They also found that citizens' perception of the accuracy of e-government services had no significant effect on their satisfaction. Moreover, in a study of citizen users' satisfaction with e-government services in general, Lee (2011) reported that satisfaction was significantly related to three dimensions of e-government service: information quality, information security, and e-participation quality. The study also found that individual IT capability is not a significant factor affecting satisfaction.

Table 5.6 E-Government Effects on Organizational Output: Multivariate Studies

Dimension	Dependent variable	Positive impact	Insignificant impact	Research method	Reference
User satisfaction	Satisfaction with KMS	Quality knowledge, KMS quality		2008 survey of 2,275 employees at 4 public institutions	Lee (2010)
	Satisfaction with e-government services	Responsiveness; speediness; openness to communication; convenience; reliability.	Accuracy	2004 survey of 485 citizens who used local e-government services (Daegu)	Sung and Jang (2005)
		information quality; information security; e-participation quality	IT capability	2008 survey of 1,214 citizens	Lee (2011)
Effectiveness	Perceived effectiveness of e-government services	Communication networks among IT, program units, and IT vendors		Survey of 174 officials at 2 district governments (Gangnam and Seocho) in 2005 and 2006	Lee (2013)

Note: IT = information technology; KMS = knowledge management system.

Economic Impacts

The review of studies on the economic impact of Korean e-government revealed five dimensions of impact: cost reduction, return on investment, employment, local economy, and globalization.

Cost Reduction

The relationship between e-government and cost reduction is not strongly supported by e-government supply-side research (Moon 2002), though demand-side research reveals that e-government users, such as citizens and businesses, do gain cost-reduction benefits. However, e-government research in Korea tells a different story (table 5.7). According to one early e-government research study (Yu, Kim, and Yoo 1994), 61 percent of government officials in central agencies reported in 1994 that computer use in their organization led to cost savings; 91 percent indicated that computer use led to time savings to complete work. At the local government level Han (2005) reported similar results; 62 percent of local government officials experienced time reductions to complete tasks due to greater IT investment by their local governments.

These descriptive findings are supported by a multivariate study (Lee 2010) that showed that employees' use of KMS is positively related to perceived cost and time savings. But it should be noted that only 22 percent of government officials at central agencies experienced a decreased workload, while 36 percent reported an increased workload due to computer use (Yu, Kim, and Yoo 1994). In addition to cost reduction, Han (2005) found that 64 percent of officials of local governments confirmed that local government IT investment reduced wait time for citizens who needed government services.

Return on Investment and Employment

Scholars in the field of MIS have researched the return on investment of IT and have applied economic models to examine the impacts of IT investment on

Table 5.7 E-Government Effects on Cost Reduction: Descriptive Studies

Dimension	Dependent variable(s)	ICT applications	Positive response (%)	Negative response (%)	Research method	Reference
Cost savings	Cost savings	Computer use in general	61	10	Survey of 422 officials at 10 central agencies	Yu, Kim, and Yoo (1994)
Increased workload	Increased workload		36	22		
Reduced time	Saving time to complete task		91	3		
	Reduced time to complete tasks	General IT investment	62	5	2003 survey of 552 officials at three local governments	Han (2005)
	Reduced wait time for citizens		64	7		

Note: ICT = information and communication technology; IT = information technology.

Bringing Government into the 21st Century • http://dx.doi.org/10.1596/978-1-4648-0881-4

economic gains (e.g., Lee and Perry 2000). Some studies (e.g., Carr 2003) have offered pessimistic views of the return on IT investment, while others (Lee and Perry 2000) believe that IT investment matters for economic gains.

Because of the financial crisis during the late 1990s, Korea suffered from severe unemployment, and the government faced strong demands to create more jobs. In response to these social demands, the government actively adopted e-government as a strategic means of reforming government bureaucracy, boosting the national economy by facilitating the IT industry, and creating jobs through government-led IT projects. One such effort, the IT New Deal Projects, was designed to hire, train, and help people get jobs using IT skills, including digitizing and building government-wide databases. As shown in table 5.8, Lee, Park, and Ju (2000) conducted a cost-benefit analysis by focusing on direct measures and found that IT investment for the building of government-wide databases (e.g., labor costs) in 15 central government agencies generated positive benefits (e.g., employment and digitization) ranging from $0.9 million to $20 million for these agencies.

In a similar vein, one government report (Korea Communication Commission [KCC] 2000) used direct measures to analyze the impact of investing in 32 IT New Deal Projects affecting 21 central government agencies from 1999 to 2000. The analysis found that (a) the $12 billion invested generated $65 billion in economic output; (b) the projects employed on average 11,116 persons per day; and the revenue of 57 medium and small IT vendors increased. Other scholars (Park, Ju, and Choi 2002) examined the relationship between the IT budget at

Table 5.8 E-government Effects on Return on Investment and Employment

Dependent variable(s)	ICT applications	Findings	Research method	Reference
Benefits (employment and digitization)	Costs (e.g., labor) for building databases	Benefits range from $0.9 million to $20 million	Analysis of 15 central government agencies	Lee, Park, and Ju (2000)
Cost benefits; employment; medium/small business	Costs of IT projects (e.g., labor)	$65 billion; 11,116 employed per day on average; 57 medium and small size; IT vendors' revenue increase	Analysis of 32 IT New Deal Projects (digitization projects) in 21 central government agencies	Korea Communication Commission (2000)
Employment	$30 million IT budget between 1998 and 2000	Creation of 48,019 jobs in industries related to 7 central agencies	Survey of seven central government agencies	Park, Ju, and Choi (2002)
Perceived usefulness of getting a job	IT New Deal Projects	61% (positive response); 18% (negative response)	Survey of 119 individuals who got jobs after the project participation	Lim and Park (2002)
Getting a job	Satisfaction with IT New Deal Projects participation	Insignificant	2001 survey of 311 individual participants in 4 central government agencies IT New Deal Projects	

Note: ICT = information and communication technology; IT = information technology.

central government agencies and the creation of industry jobs. They reported that the $30 million that constituted the IT budgets at seven agencies created 48,019 jobs in industries related to these agencies. Using a 2001 survey of 311 participants in IT New Deal Projects, Lim and Park (2002) found that 61 percent of participants believed that project participation would be useful in landing a job. However, this study also reported that only 119 participants (38.3 percent) got jobs after their project participation. Unlike the original policy expectation, the results of logistic analysis revealed that participants' satisfaction with the IT New Deal Projects had no significant impact on getting a job.

Local Economy and Globalization

As shown in table 5.9, two descriptive studies reported on e-government effects on local economy and globalization. Jung and Son (2007) studied the effects of the Digital Village Project—a government-driven Informatization project to boost the economy and build communities in rural areas in Korea. Using a 2006 survey of 390 local residents in six communities in a state-level local government (Kyungsang-Buk Do), this study documented that 37 percent of respondents agreed that the Digital Village project had had positive effects on their local economy, and 30 percent disagreed with this statement. Also, the authors reported that, overall, early adopters in 2001 tended to have more positive perceptions toward local economy impacts of the Digital Village program than later adopters in 2002 and 2003.

Advancement of ICT infrastructure and e-government applications in Korea has provided local governments with the opportunity to reach out to global communities and to build and strengthen formal and informal relationships with them by seeking, gaining, and exchanging information and by doing business electronically. Lee and Lee (2002) studied how e-government applications improve local governments' globalization programs, such as

Table 5.9 E-Government Effects on Local Economy and Globalization

Dependent variable(s)	ICT applications	Positive response (%)	Negative response (%)	Research method	Reference
Local economy	Digital Village	37	30	2006 survey of 390 citizens from six communities in Kyungsang-Buk Do	Jung and Son (2007)
Improvement of globalization programs in local government	E-government adoption	74–93	0–9	2001 survey of 93 local governments	Lee and Lee (2002)
	E-documents	50–76	0–9		
	Information sharing	63–83	0–8		
	ICT infrastructure	54–93	0–23		
	E-civil applications	46–73	0–24		
	Databases	67–100	0–16		

Note: ICT = information and communication technology.

Bringing Government into the 21st Century · http://dx.doi.org/10.1596/978-1-4648-0881-4

sister/brother local governments in other countries, tourism promotion, and cultural and economic exchange programs. Using a 2001 survey of 93 local governments, Lee and Lee found that e-government adoption, e-documents, information sharing, and database systems improved local government's globalization programs.

Political Impacts

Drawing on Rosenbloom's (1983) political value of public administration, this assessment classified research on the political impacts of e-government in Korea into two broad dimensions: responsiveness and accountability.

Responsiveness

Table 5.10 shows the effects of Korean e-government on different aspects of responsiveness: response rate and time, and the use and quality of citizens' online input in the policy-making process.

Response rate and time. Lee and Min (2002) investigated how the adoption of a new e-participation policy, called Internet Real-Name systems, in the city of

Table 5.10 E-Government Impacts on Responsiveness

Dimension	Effect	Key findings	Research method	Reference
Response rate and time	Response rate and time	Adoption of Internet Real-Name systems increased response rate and decreased response time	A content analysis of Jinjoo City websites (Open Mayor Office and Online Forum)	Lee and Min (2002)
Use and quality of citizen Input	Use of citizen input to make environmental policy decisions	*Significant factors*: e-government service quality, e-participation services, and top management support; *Insignificant factors*: political importance of environmental issues, environmental activities in the community, perceived seriousness of environmental issues in the community	2003 survey of 315 city officials at 74 environment agencies in city government; content analysis of those city government websites	Lim (2006)
	Quality of citizens' online input	*Significant factors*: environmental activism, IT leadership, e-government service quality; *Insignificant factors*: environmental pollution, political importance of environmental issues	2003 survey data from 315 city officials at 74 environment agencies in city government	Lim and Tang (2008)
	Use of online citizen survey results	*Significant factors*: policy agenda-setting and formulation stages, soliciting citizens' support ("yes" or "no"); *Insignificant factors*: policy implementation and evaluation stages, asking citizens' preferences	Content analysis of 434 online survey results available at Gangnam e-government	Ha and Park (2008)
	Use of online survey results	Tension with local council and increased workload	Interviews with 45 local officials in Gangnam district, Seoul	Ahn and Bretschneider (2011)

Note: IT = information technology.

Jinjoo in July 2001 affected the local government's responses to citizens' input posted through e-participation programs. By analyzing two e-participation programs—Open Mayor Office and Open Forum—before and after the new e-participation policy, spanning 22 months, they found that the response rate had significantly increased to 99.41 percent from 35.24 percent, and response time had been reduced from an average of 3.44 days to 2.77 days after the adoption of the new e-participation policy. Also reported was a decrease in the number of postings (both inquiries and complaints) via these two e-participation programs after the adoption of the new policy.

Citizen input. Lim (2006) examined the factors affecting the extent to which city government used e-participants' input to make decisions on environmental policy. Analyzing a 2003 survey of 315 city officials at 74 environment agencies, Lee found that e-government service quality, e-participation service quality, and top management support were significantly related to the use of e-participants' input. On the other hand, city officials' perceptions of the political importance of environmental issues, environmental activities, and the seriousness of environmental issues in the community had no significant relationship. The findings implied that the high quality of e-government and e-participation service affected city officials' response to e-participation. In other words, city officials' use of e-participants' input is primarily driven by internal factors, not external forces.

Using the same data, Lim and Tang (2008) found that environmental activism, IT leadership, and e-government service quality are significantly related to the quality of e-participants' input, but the seriousness of environmental pollution and the political importance of environmental issues are not.

Ha and Park (2008) examined how different stages of the policy-making process and government intentions to seek e-participants' input affect the use of online survey results in policy making. By analyzing 434 online survey results available at the Gangnam district's websites, they found that a local government uses the results of an online citizen survey to make policy decisions when the survey is an online poll designed to gain citizens' support ("yes" or "no" choice only) during the policy formulation stage. In other words, the findings imply that the results of an online survey are less likely to affect policy decisions when the survey is designed to identify citizens' preferences or to ask questions about policy implementation and evaluation. Ahn and Bretschneider (2011) studied the same local government by conducting interviews with 45 local officials in 2005. They found that the Gangnam district actually used the results of an online survey to make policy decisions. The interviewees also said that the use of online survey results often increased local officials' workload and at times created conflict with local council members.

Accountability

Advocates have viewed e-government as a means of improving government accountability. It is believed that one way of ensuring accountability is to enhance transparency by making government information available and accessible to the

Table 5.11 E-Government Impacts on Accountability

Dimension	ICT applications	Key findings	Research method	Reference
Transparency in business processes	IT investment	Positive: 58% Negative: 7%	2003 survey of 552 officials at three local governments	Han (2005)
Corruption and integrity	Open systems use	Positive, but increased workload	Interviews with 15 officials at SMG in 2006 and 2007	Kim, Kim, and Lee (2009)
Transparency	Internet broadcasting of senior staff meeting; online publication of official documents	Positive, but increased workload; protected bureaucracy from negative media attacks	Interviews with 45 local officials in Gangnam district in 2005	Ahn and Bretschneider (2011)
Transparency	OASIS	*Significant factors*: e-participants' sense of belonging to a community and influence on city decision making; satisfaction with e-participation applications; *Insignificant factor*: satisfaction with government responsiveness	2009 survey of 1,076 citizens who used OASIS programs in SMG in 2009	Kim and Lee (2012)

Note: IT = information technology; OASIS = Organization for Advancement of Structured Information Standards; SMG = Seoul Metropolitan Government.

public (La Porte, Demchack, and de Jong 2002; Welch and Wong 2001). Scholars in e-government in Korea have paid attention to the role of e-government in improving transparency in government, and have provided some evidence to support the beliefs of the e-government advocates (table 5.11).

Han (2005) analyzed the 2003 survey data of 552 local government officials and concluded that 58 percent of respondents said the impact of IT investment (e.g., IT budget) increased transparency in business processes. Another descriptive study (Ahn and Bretschneider 2011) also reported that local government officials believed that e-government services, such as Internet broadcasting of senior staff meetings and online publication of official documents, positively affected transparency in government. They added that enhanced transparency does increase workload, but it protects the bureaucracy from negative media attacks. Using interview data from 15 Seoul Metropolitan Government (SMG) officials in 2006 and 2007, Kim, Kim, and Lee (2009) found that the SMG's efforts to enhance transparency by using open systems reduced corruption and promoted the integrity of SMG officials. Kim and Lee (2012) examined the effects of citizens' e-participation program experiences—that is, the Organization for Advancement of Structured Information Standards (OASIS) programs—on transparency in SMG. They found that e-participants perceived enhanced transparency in SMG when they believed that e-participation enabled them to be more interested in community issues, get a better sense of belonging to a community, and influence government decision making.

Social Impacts

The literature review reveals that the findings related to social impacts of e-government in Korea are classified into two broad categories: trust in government and social inclusion/cohesion.

Trust in Government

E-government's impact on trust in government has received growing attention by scholars and practitioners around the world (Kim and Lee 2012; Tolbert and Mossberger 2006; Welch, Hinnant, and Moon 2005). As e-government emerged with the creation and commercialization of Internet technologies in the late 1990s, some scholars attempted to understand the broad effects of the Internet on trust in government in Korea (Im et al. 2014). This section discusses empirical research examining the relationship between e-government and trust in government in Korea.

As table 5.12 depicts, these studies found that e-government has had a positive impact on trust in government. According to a descriptive analysis of a 2003 survey data of 552 local government officials (Han 2005), 55 percent of respondents agreed with the statement that IT investment leads to enhanced trust in their local government. The findings were supported by a study of trust in

Table 5.12 E-Government Impacts on Trust, Social Inclusion, and Cohesion

Dimension	ICT applications	Key findings	Research method	Reference
Trust in local government	IT investment	Positive: 55% of respondents Negative: 7% of respondents	2003 survey of 552 officials at three local governments	Han (2005)
Trust in government in general	E-government services	Satisfaction with e-government service and trust in e-government increases trust in government in general	Survey of 1,214 citizens in Korea	Lee (2011)
Trust in SMG	OASIS	Perceived transparency in SMG increases trust in SMG	2009 survey of 1,076 e-participants of OASIS in SMG	Kim and Lee (2012)
Reduced digital divide	Digital Village adoption	Positive: 28%; negative: 34%	2006 survey of 390 citizens from six communities in Kyungsang-Buk Do	Jung and Son (2007)
Enhanced community building		Positive: 30%; negative: 30%		
Quality of life		Positive: 29%; negative: 30%		

Note: IT = information technology; OASIS = Organization for Advancement of Structured Information Standards; SMG = Seoul Metropolitan Government.

Bringing Government into the 21st Century • http://dx.doi.org/10.1596/978-1-4648-0881-4

government by Lee (2011), who analyzed the 2008 survey data of 1,214 citizens and found that citizens' trust in government in general is positively affected when they are satisfied with e-government. Kim and Lee (2012) added evidence showing that e-participants' perception of SMG efforts to improve transparency was positively related to trust in SMG.

Social Inclusion and Cohesiveness

With the advancement of Internet technologies, digital-divide issues have become a concern among government leaders, community leaders, and e-government scholars (e.g., Brown and Garson 2013). The Korean government has attempted to address the digital divide in Korea that excludes rural communities from the advantages of technology, including e-government. Government officials have been concerned that some people in rural communities are isolated and have limited opportunities to interact with community members and receive public services. It is argued that this limitation stems in part from the lack of IT skills and Internet connectivity required to access and share information and to receive public services, such as education, health, and leisure, that are provided by e-government technologies.

To resolve digital-divide issues and strengthen social cohesion in rural communities, the government launched the Digital Village projects in 2001. These projects are designed to address digital literacy and to improve social inclusion and cohesion. As shown in table 5.12, Jung and Son (2007) assessed the impact of Digital Village projects on social inclusion and cohesion in pilot communities. According to a descriptive analysis of a 2006 survey of 390 citizens of six pilot communities in a state-level government (Kyungsang-Buk Do), respondents reported that social inclusion and cohesion impacts of Digital Village projects were not yet realized. For example, only 28 percent of respondents believed that the Digital Village project reduced the digital divide, whereas 34 percent did not believe this was the case. Approximately 30 percent of respondents reported positive impacts of programs to counter the digital divide, with a similar percentage reporting negative impacts. These impacts related to enhanced social cohesion (e.g., information sharing with other community members) and quality of life (e.g., education, health, leisure). However, this research also found that respondents in early adopter communities (since 2001) had more positive perceptions toward the impacts of the Digital Village program on social inclusion and cohesion than those in later adopter communities (since 2002 or 2003).

Conclusions

Lessons Learned about Administrative Impacts

Positive administrative impacts learned from the descriptive analysis studies include reduced red tape, strengthened formalization, expanded span of control of middle managers at central agencies and local governments, better decision making in terms of policy goal-setting and searching for policy alternatives, improved business processes and administrative efficiency, and increased end-users' satisfaction with the outcomes of e-government systems and services.

Multivariate analysis research provides further information about the positive administrative impacts of e-government and how they can be achieved.

More effective policy decisions. The findings imply that government officials' policy decisions can be more effective when government provides high quality e-government websites and uses citizens' input secured from these online sites. Furthermore, policy decisions can be enhanced when government officials are supported by IT leadership and the use of the diverse functions of information systems. As briefly discussed earlier, the Korean government has initiated Government 3.0 by emphasizing the use of IT, especially big data, to make government smarter and more competent. A recent study (Lee 2015) found that the use of big data helps government make better policy decisions in the context of emergency management by combining new data about the trends of social interest, such as the emergence and diffusion of Influenza A (H1N1), and conventional information collected through emergency management processes.

Better quality and use of information. Communication and business processes can be improved when internal e-government systems are designed and implemented to increase convenience, generate high quality information, and ensure information security. In addition, e-government systems can produce high quality information and knowledge when leaders have greater interest in and support for e-government systems, such as KMS. E-government systems are most valuable when end-users use the outputs of the systems. The assessment also reveals that the use of knowledge generated by KMS can be improved when government officials have greater absorptive capacity, self-efficacy and perception of the usefulness of knowledge, and greater interpersonal trust with their co-workers.

Increased satisfaction. Korean e-government research implies that government officials' satisfaction with e-government systems is likely to increase when the systems are easy to use, useful, and produce high quality information. In a similar vein, citizens are satisfied with e-government services when government uses e-government to appropriately respond to their needs, engage with them, and provide secure and reliable information and services.

However, research indicates that some administrative impacts of e-government in Korea have been rather limited. Downsizing effects are observed at central agencies, but not at local governments. Although e-government systems such as KMS are designed to provide government officials with greater access and ease of use of information, these benefits have been limited to knowledge created by KMS. In addition, multiple factors such as rewards, the culture of organizational learning, and managerial support do not appear to be significant factors related to knowledge use, which suggests that managerial and organizational factors are not always significant in affecting organizational behavior related to the use of IT-generated information.

Given the lessons learned about e-government's impact in Korea on administration, the following recommendations are provided to e-government leaders, managers, and designers.[1] E-government leaders should:

- Build IT leadership skills and capabilities,
- Commit to e-government by stressing the value of e-government systems,
- Support the adoption and use of e-government systems,
- Embrace the value of interpersonal trust and make efforts to build a culture of interpersonal trust in the organization.

E-government managers should:

- Be trained to effectively manage the expanded span of control,
- Encourage employees to use the diverse functions within e-government systems to make better decisions,
- Be trained to cultivate and enhance their ability to explore information outside siloed work units and organizational structures.

E-government designers, such as chief information officers and IT managers in developing countries, should:

- Design and implement e-government systems that improve convenience and provide secure, reliable, and useful information and knowledge,
- Engage end-users, such as non-IT government officials and citizens, when designing e-government systems.

Lessons Learned about Economic Impacts

Research on the economic impacts of e-government in Korea reveal mixed findings. Among the five economic impact dimensions, it seems that e-government in Korea has fostered *cost reduction, return on investment*, and *improvement of relationships with the international community*. It appears that e-government has enabled government officials to save time and costs when performing their work; and it has also enabled citizens to save time when doing business with the government.

In addition, the government's 32 IT New Deal Projects seemed to generate a fair return on investment in terms of overall cost/benefit at central agencies. The results were particularly strong for projects that enabled central agencies to build databases and information systems for diverse functional domains, such as an e-procurement electronic data interchange (EDI) database and public health information systems that facilitated the speed of digitization of government data previously stored in analog format. Along with the establishment of the telecommunication network backbone in the early 1990s, digitization, as a basic building block of e-government's technological platform, appears to have equipped the Korean government with a core capability of advancing innovative e-government services.

Along with e-government's impact on cost reduction of work processes, e-government also resulted in increased workloads for government officials (Ahn and Bretschneider 2011; Yu, Kim, and Yoo 1994). The reason may be the lack of re-engineering efforts to integrate the use of ICT with existing business processes (Hammer and Champy 1993). For example, some studies observed that government officials use electronic approval systems to draft and send out a document to get approval from their bosses, but they also bring a hard copy of the document to their bosses in person (Lee 2008).

The literature review also indicated, however, that e-government impacts on *employment* and the *local economy* were weaker than expected. In terms of jobs, digitization projects seem to have created employment opportunities and satisfied participants who sought employment, but we have also learned that participants' satisfaction with the projects did not improve their ability to obtain work. A lesson learned from these observations is that when e-government projects are designed to create employment by hiring people to work for the e-government projects, and also to train individuals to get jobs after the projects are completed, e-government projects should be carefully planned and implemented by collaborating with private IT vendors who are potential employers of project participants. Other lessons learned from a study of Digital Village impacts are that overall e-government impacts on local economies have not met expectations in rural Korean communities. We also see that early community adopters of e-government projects, such as Digital Village, gain significantly better economic benefits than late adopters and that e-government impacts on the local economy can be observed only with the earlier adopters.

Based on these findings, we provide recommendations for the key roles of e-government leader, manager, and designer.

E-government leaders should:

- Embrace the value and impact of digitizing government information for building e-government as a means of reducing administrative costs and boosting the national economy,
- Set short-term and long-term goals and timelines for government-wide e-government projects,
- Provide support for IT investment,
- Embrace e-government as an effective means of building global networks and economic development partners with businesses and governments of other countries.

E-government managers should:

- Streamline and re-engineer business processes prior to adopting e-government applications or while implementing them (if they are already adopted) to minimize the increased workload issue,
- Facilitate the use of e-government applications to improve globalization programs that will build and strengthen relationships with other countries

E-government designers, such as chief information officers and IT managers in developing countries, should:

- Use their IT expertise to design and implement the digitization of government information and develop government-wide databases and e-government systems,
- Build strategic and collaborative relationships with IT vendors as potential employers, as well as IT experts, and engage them in government-wide IT projects designed to facilitate employment.

Lessons Learned about Political and Social Impacts

Overall, e-government in Korea has had a positive impact on the two political dimensions of responsiveness and accountability.

Responsiveness. Government response to citizens' demands has been observed in the area of e-participation at the local government level. The findings suggest that the new e-participation policy, Real-Name systems, has motivated local governments to respond to online citizens' input in a timely manner. More importantly, local government has responded to citizens' needs by listening to e-participants' voices and using citizen input to make decisions. Use of e-participants' input is facilitated by top management support, providing high quality e-government and e-participation services. Critics of citizen participation have emphasized the importance of educating citizens because citizens lack the knowledge to provide useful input about public policies (e.g., Dahl 1989). The findings, however, suggest that city officials perceive e-participants' input to be of higher quality when greater environmental activism is apparent in the community, when the city provides high quality e-government services, and when e-participation is supported by city IT leadership.

The findings also suggest that a local executive body actively uses e-participants' input as a means of initiating the setting and formulation of a policy agenda; however, this activity can create conflict with a local legislative body (Ahn and Bretschneider 2011). The tension between executive and legislative bodies seems to arise when council members, as representatives of public interest, believe that the executive body abuses the results of online citizen surveys as a means of justifying policy decisions and gaining budget approval to implement policies. With regard to the use of e-participation input, the tension between executive and legislative bodies is consistent with ongoing debates in conventional citizen participation literature (Roberts 2004). This issue becomes critical when executive and legislative bodies differ on a particular policy issue.

Accountability. In terms of e-government's impact on transparency, the literature seems to provide consistent findings that e-government improves transparency in government. The findings suggest that transparency in government can be enhanced by greater IT investment and by specific e-government applications designed to minimize the possibility of corruption. Although the anticorruption index and Integrity Assessment scores [2] have increased since the OPEN system was adopted in 1999, it should be noted that some studies (e.g., Choi 2007) argue that decreased corruption in SMG is probably not because of the adoption of the OPEN system, but because of an overall decrease in corruption in the Korean government. The findings also imply that e-participation programs, such as OASIS in SMG, positively affect transparency in government when these programs meet e-participants' expectations about development value and political efficacy.

Social impacts. The selected studies on the social impacts of Korean e-government show mixed findings. Although e-government appears to have had a positive influence on trust in the Korean government, social inclusion and cohesion have not been positively observed. The findings suggest that government's IT investment and citizens' perception of transparency in government matters for building trust in government at the local government level. Another finding implies that government can restore citizens' trust in government by providing high quality e-government and e-participation services to satisfy citizens' expectations. The study of Digital Village's impact on social inclusion and cohesion, however, suggests that it will take time to bridge the digital gap and build social cohesion in underdeveloped communities in terms of IT infrastructure and human resources. But this study also implies that social inclusion and cohesion can be improved when rural communities are exposed to emerging e-government technologies as early as possible.

The political and social impacts of e-government in Korea have been discussed here in the context of local e-government. It should be noted that the lessons about these impacts might not be equally applicable to central e-government in developing countries. Therefore, the following recommendations are probably more relevant to local e-government in developing countries.

E-government leaders should:

- Provide support for listening to citizens' input via e-participation,
- Collaborate with a legislative body to understand citizens' online input and collaborate with citizens to make more democratic and effective policy decisions,
- Embrace the value of e-government, e-participation services, and transparency as a means of restoring trust in government,
- Pay more attention to the speed of e-government diffusion to rural communities.

E-government managers should:

- View citizens as collaborators (not only customers) of e-governance development,
- Sincerely respond to citizens' input via e-participation and let citizen participants know that their voices are heard,
- Develop e-government policies that will open government information to the public to enhance transparency in government.

E-government designers should:

- Design, build, and provide high quality e-government and e-participation services to engage citizens online,
- Design and implement the e-participation applications that provide useful information related to government policy and easy-to-use functions,
- Design and implement e-government and e-participation systems that meet citizens' expectations about e-government services and enhance transparency.

Notes

1. The idea of three roles around e-government was offered by Professor Richard Heeks, who served as a peer reviewer for this publication.
2. The Anti-Corruption and Civil Right Commission in Korea has periodically conducted a citizen survey to assess the integrity of public sector organizations and released the results to the public. Public sector organizations include, but are not limited to, central government agencies, local governments, public hospitals and public schools.

Bibliography

Ahn, M., and S. Bretschneider. 2011. "Politics of E-Government: E-Government and the Political Control of Bureaucracy." *Public Administration Review* 71 (3): 405–13.

Andersen, K., H. Henriksen, R. Medaglia, J. Danziger, M. Sannarnes, and M. Enemarke. 2010. "Fads and Facts of E-Government: A Review of Impacts of E-Government (2003–2009)." *International Journal of Public Administration* 33: 564–79.

Bhatnagar, S. 2003. "The Economic and Social Impact of E-Government." A Background Technical Paper for the Proposed UNDESA Publication, E-Government, the Citizen and the State: Debating Governance in the Information Age. http://www.iimahd .ernet.in/~subhash/pdfs/UNDESAeGovReport.pdf.

Bretschneider, S., and I. Mergel. 2010. "Technology and Public Management Information Systems: Where Have We Been and Where Are We Going". In *The State of Public Administration: Issues, Problems and Challenges*, edited by D.C. Menzel and H. L. White, 187–203. New York: M. E. Sharpe.

Brown, M., and D. Garson. 2013. *Public Information Technology and E-Governance: Managing the Virtual State*. Hershey, PA: IGI Global.

Carr, N. 2003. "IT Doesn't Matter." *Harvard Business Review* (May): 41–49.

Choi, J. 2007. "The Impact of OPEN system on Corruption." *Governmental Studies* 13 (1): 215–40.

Choi, H., and S. Hahn. 2008. "Interaction of Information Technology and Organizational Restructuring Strategies: Cases of TIS and HTS in the National Tax Service." *Korean Public Administration Review* 42 (1): 323–44.

Chung, C. 2009. *The Theory of Electronic Government*. Seoul: Seoul Economy and Business.

Dahl, R. 1989. *Democracy and Its Critics*. New Haven, CT: Yale University Press.

Eom, S., and B. Kim. 2005. "An Empirical Study on the Influence of Public Infomatization on the Number of Employees and Middle Management in Korean Local Governments." *Korean Journal of Public Administration* 14 (3): 155–84.

Ha, H., and C. Park. 2008. "How Does Local Government Use On-Line Citizen Participation in the Policy Process? The Case of On-Line Survey of Gangnam-Gu, Seoul." *Korean Policy Studies Review* 17 (2): 93–118.

Hammer, M., and J. Champy. 1993. *Reengineering the Corporation: A Manifesto for Business Revolution*. New York: HarperCollins.

Han, S. 2005. "An Empirical Study on the Perception of the Information Technology Investment Effects in the Public Sector." *Korean Public Administration Review* 39 (1): 237–59.

Im, T. 2011. "Information Technology and Organizational Morphology: The Case of the Korean Central Government." *Public Administration Review* 66 (1): 435–43.

Im, T., W. Cho, G. Porumbescu, and J. Park. 2014. "Internet, Trust in Government, and Citizen Compliance." *Journal of Public Administration Research and Theory* 24: 741–63.

Jung, W., and N. Son. 2007. "A Study on the Performance Evaluation of the Information Network Village." *Journal of Korean Association for Regional Information Society* 10 (3): 19–43.

Kim, G. 2004. "The Key Factors Affecting Successful Knowledge Transfer in Public Officials" *Korean Public Administration Review* 38 (1): 45–68.

Kim, S., H. Kim, and H. Lee. 2009. "An Institutional Analysis of an E-government System for Anti-Corruption: The Case of OPEN" *Government Information Quarterly* 26: 42–50.

Kim, S., and H. Lee. 2006. "The Impact of Organizational Context and Information Technology on Employee Knowledge-Sharing Capabilities." *Public Administration Review* 66 (3): 370–85.

Kim, S., and J. Lee. 2012. "E-Participation, Transparency, and Trust in Local Government." *Public Administration Review* 72 (6): 819–28.

Korea Communication Commission (formerly Ministry of Information and Communication). 2000. "The 3rd Assessment of IT New Deal Projects."

La Porte, T., C. Demchak, and M. de Jong. 2002. "Democracy and Bureaucracy in the Age of the Web: Empirical Findings and Theoretical Speculations." *Administration and Society* 34: 411–46.

Lee, E., 2015. "Big Data and Policy Decision: An Application of Social Interest on Crisis Management." *Korean Policy Studies Review* 24 (4): 491–515.

Lee, G., and J. Perry. 2000. "Are Computers Boosting Productivity?" *Journal of Public Administration Research and Theory* 12 (1): 77–102.

Lee, H. 2010. "A Study on the Information System and Performance in the Public Sector: With Focus on the Public Enterprise's KMS." *Korean Policy Studies Review* 19 (2): 275–305.

Lee, J. 2008. "Determinants of Government Bureaucrats' New PMIS Acceptance: The Role of Organizational Power, IT Capability, Administrative Role and Attitude." *The American Review of Public Administration* 38 (2): 180–202.

Lee, J. 2013. "Exploring the Role of Knowledge Networks in Perceived E-Government: A Comparative Case Study of Two Local Governments in Korea." *American Review of Public Administration* 43 (1): 89–108.

Lee, J., K. Park, and H. Ju. 2000. "Investment Effect Analysis of Public Work on Informatization." *Korean Policy Studies Review* 9 (3): 217–35.

Lee, S. 2011. "A Study on Relations between Trust in E-Government and Trust in Government: Focused on the Factors of Truster and Trustee." *Informatization Policy* 18 (2): 49–71.

Lee, S., and B. Min. 2002. "The Effects of Adopting the Real Name System for Citizen Participation in Websites of Local Governments." *Korean Public Administration Review* 36 (2): 205–29.

Lee, Y., and K. Lee. 2002. "An Impact Analysis of the E-Government Program on Local Governments' Internationalization Programs in Korea." *Journal of Korean Association for Regional Information Society* 5 (2): 129–56.

Lee, Y., and C. Oh. 2000. "An Empirical Study of the Relationship between Administrative Informatization and Productivity in the Wide Area Local Government." *Korean Policy Studies Review* 9 (1): 163–89.

Lim, D., and G. Park. 2002. "An Empirical Analysis of Employment Effect on the Information Technology New Deal." *Korean Policy Studies Review* 11 (2): 165–94.

Lim, J. 2006. "Urban E-Government and Citizen Participation in Environmental Governance." *Korean Public Administration Review* 40 (3): 53–76.

Lim, J., and Y. Kang. 2013. "E-Government Systems and Decision-Making Performance in the Public Sector." *Journal of Korean Association for Regional Information Society* 16 (4): 27–54.

Lim, J., and S. Tang. 2008. "Urban E-government Initiatives and Environmental Decision Performance in Korea." *Journal of Public Administration Research and Theory* 18 (1): 109–38.

Mok, J., Y. Choi, and S. Myeong. 1998. "The Impact of Information Technology in Policy-Making Process: The Case of Korean Local Governments." *Korean Public Administration Review* 32 (3): 35–54.

Mok, J., S. Myeong, and T. Yun. 2002. "Reduction of Administrative Corruption by E-Government: Focusing on Administrative Red-Tapes and Information-Communication Technology." *Informatization Policy* 9 (3): 3–17.

Moon, M. 2002. "The evolution of E-government among Municipalities: Rhetoric or Reality?" *Public Administrative Review* 62: 424–33.

Moon, M., J. Lee and C. Roh. 2014. "The Evolution of Internal IT Applications and E-government Studies in the Public Administration Discipline: Research Themes and Methods." *Administration & Society* 46 (1): 3–36.

Morgeson III, F. 2012. "Expectations, Disconfirmation, and Citizen Satisfaction with the US Federal Government: Testing and Expanding the Model." *Journal of Public Administration Research and Theory* 23 (2): 289–305.

Myeong, S., and Y. Choi. 2010. "Effects of Information Technology on Policy Decision-Making Processes: Some Evidences Beyond Rhetoric." *Administration & Society* 42 (4): 441–59.

Park, K., H. Ju, and B. Choi. 2002. "Estimating Employment Induced Effects Produced by the IT New Deal Program: An Application of the Input-Output Analysis." *Korean Policy Studies Review* 11 (3): 89–113.

Rainey, H. 2014. *Understanding and Managing Public Organizations.* San Francisco, CA: Jossey-Bass.

Roberts, N. 2004. "Public Deliberation in an Age of Direct Citizen Participation." *American Review of Public Administration* 34 (4): 315–53.

Rosenbloom, D. 1983. "Public Administrative Theory and the Separation of Power." *Public Administration Review* 43: 219–26.

Rosenbloom, D., R. Kravchuk, and R. Clerkin. 2009. *Public Administration: Understanding Management, Politics, and Law in the Public Sector.* New York: McGraw-Hill.

Shim, D., and T. Eom. 2009. "Anticorruption Effects of Information Communication and Technology and Social Capital." *International Review of Administrative Sciences* 75 (1): 99–116.

Song, H., T. Cho, H. Kwon, H. Yu, and S. Kim. 2008. "Analysis of the Determinants of the Performance of E-Government Policy: Focusing on the National Education Information System." *Korean Policy Studies Review* 17 (4): 223–48.

Sung, D., and C. Jang. 2005. "Evaluation of Customer-Orientedness of Public Services: Focused on the e-Government. *Korean Public Administration Review* 39 (2): 207–32.

Tolbert, C., and K. Mossberger. 2006. "The Effect of E-Government on Trust and Confidence in Government." *Public Administration Review* 66 (3): 354–69.

Welch, E., C. Hinnant, and M. Moon. 2005. "Linking Citizen Satisfaction with E-Government and Trust in Government." *Journal of Public Administration Research and Theory* 15 (3): 371–91.

Welch, E., and W. Wong. 2001. "Global Information Technology Pressure and Government Accountability: The Mediating Effect of Domestic Context on Website Openness." *Journal of Public Administration Research and Theory* 11 (4): 509–38.

Yu, P., D. Kim, and B. Yoo. 1994. "An Empirical Analysis of the Employees' Perceived Impacts of Computerization upon Public Organizations." *Korean Public Administration Review* 28 (4): 1371–86.

Yun, S. 2000. "An Analysis of Informatization Impact on National Competitiveness." *Korean Public Administration Review* 34 (3): 47–71.

CHAPTER 6

Lessons and Implications for Developing Countries

Tina George Karippacheril

Introduction

The Republic of Korea's journey from a developing country in the 1960s to an advanced information society in the 21st century has been remarkable. Investments in technology were ramped up from the early 1970s under military leadership, through the late 1980s, as it transitioned to a democracy and a digital economy. Today, Korea is a recognized leader in several dimensions of governance in the digital age, including openness, quality of institutions, rule of law, a high-performing and competent bureaucracy, political stability, commitment of the political leadership to reforms, market institutions, commitment to inclusive development, citizen engagement, and establishment of trust between the citizen and the state. Indeed, Korea has maintained its number one ranking on the ICT Development Index (IDI) from 2010 to 2015.[1]

Korea's experience in attaining its current leadership role as an information society provides a valuable learning resource for countries at different stages of their Digital Governance initiatives. In this concluding chapter, we review the impact of Korea's Digital Governance program, and distill seven key lessons, examining how these lessons apply to two groups of countries: (A) developing (low income) countries in need of urgent support to initiate e-Government programs (including countries with an extreme poverty rate above 40%); and (B) more advanced economies (middle income) moving from fragmented information systems to connected platforms. For each lesson, we delve into the history, background and sequencing of the reforms, and draw policy implications and critical success factors. This is followed by consideration of the lessons Korea learned from the measures that did not work and how it addressed these setbacks. Given that certain technologies are expensive or not particularly efficient, we examine

With contributions from Grace P. Morgan, Robert P. Beschel, Jane Treadwell, Cem Dener and Jeongwon Yoon opportunities for leapfrogging such technologies to accelerate development. Finally, we conclude with a summary of implications from the Korean experience that can provide valuable guidance to aspiring countries.

Impacts of Digital Governance in Korea

Is there sufficient evidence on the impacts of Korea's digital governance to brand it a success? If so, what are the drivers of that success? Chapter 5 examines the available literature, albeit somewhat scarce, on four kinds of impacts—administrative, economic, political and social.

Administrative impact. The use of ICT in government has been found to have had a limited but positive impact. Specifically, the use of ICT has reduced red tape in government (Mok, Myeong, and Yun 2002) and had an impact on organizational structure. For instance, the adoption of the Home Tax Service system[2] has reinforced management's decision-making authority, contributed to the formalization of organizational policies, procedures and rules, and empowered frontline officials (Choi and Hahn 2008). On the other hand, contrary to popular perception, ICT adoption did not contribute to downsizing the number of full-time employees in local government (Eom and Kim 2005). ICT, in fact, increased the size of 10 central agencies between 1989 and 2005 (Im 2011). The span of control also increased, which meant that managers were responsible for managing more staff, adding to the managerial workload (Eom and Kim 2005; Im 2011). At both central and local levels, ICT use improved the ability of officials to make decisions (Myeong and Choi 2010), and simplified and sped up government processes (Han 2005). Moreover, citizen satisfaction was impacted by government responsiveness, speediness, openness to communication and reliability, which, in turn, were improved by the implementation of ICT in government (Sung and Jang 2005).

Economic impact. Evidence that the use of ICT has brought any cost reduction for the government is scarce. However, ICT use did reduce the time taken by local government officials to complete tasks (Han 2005), and citizens and businesses benefited from a reduction in transaction costs and wait times as result of government ICT investments (Han 2005). One of the most significant impacts of ICT investments by the government in Korea has been on job creation. Following the Asian Financial Crisis in 1997, the government placed a special emphasis on e-government projects, applying process innovation to reform bureaucratic public administration. The campaign theme was "*we may be behind in industrialization but let's lead in informatization.*" (Song 2007). Investments in the ICT industry were employed as a means of bolstering the national economy and creating jobs that serviced government contracts through a number of IT New Deal Projects.[3] Both independent and government-led studies found that ICT investments had generated substantial benefits in terms of job creation and economic output. However, one perception study has indicated that there was no assurance of employment for project participants once these projects ended, suggesting the importance of stronger collaboration with IT vendors, as potential employers of the participants, during project design and implementation.

Political impact. ICT investments in Korea have improved government responsiveness and accountability, particularly at the local level. A study on the impact of introducing two e-participation programs (Open Mayor Office and Open Forum) for a local government found that the response rate had increased from 35% to 99%, and response time had decreased from 3.4 days to 2.7 days on average (Lee and Min 2002). Furthermore, ICT investments have had an impact on improving transparency by making government information and data available to and accessible by citizens. Overall, transparent and open systems were perceived to reduce corruption and strengthen the integrity of local government officials (Kim, Kim, and Lee 2009).

Social impact. ICT has been perceived to have positively impacted citizen trust in their local governments (Han 2005; Lee 2011; Kim and Lee 2012). However, evidence is scarce that ICT investments have improved social inclusion and the cohesiveness of Korean society (Jung and Son 2007).

Lessons Learned: "The Korean Seven"

One of the criticisms of studies of ICT impact is a general view that "regard(s) IT as a 'good thing' for government" (Heeks and Bailur 2007). These studies characterize technology as transformative, "as if it alone would usher in a transformation of the state and as if politics and current institutions could be ignored in such a transformation" (Fountain 2001a). On the other hand, studies indicate that the role of ICT has hitherto been neglected by public administration scholars (Dunleavy et al. 2006). While both of these views have validity, in sum they reflect the challenge of presenting a holistic understanding of both the technological and non-technological[4] aspects of implementing digital governance programs. Indeed, this is highlighted in the World Bank's World Development Report 2016: "to maximize the **digital dividends** (broader development benefits from using digital technologies) and mitigate the risks requires better understanding of how technology interacts with the analog **complements** (other factors that are important for development: **regulations** that promote competition and entry, **skills** to leverage digital opportunities, and **institutions** that are capable and accountable)." Correspondingly, in the following core section of this chapter, we present both technological and non-technological lessons from the Korean digital governance experience that are relevant to countries undertaking these programs.

We have identified seven key lessons from the Korean digital governance experience. Box 6.1 briefly discusses these lessons, which we call "The Korean Seven," and suggests what action aspiring countries need to take in response to such insights.

Box 6.1 "The Korean Seven"

1. **Politics: Sustained high-level leadership and support for digital governance over the long term across the political spectrum provides the essential foundation for change.** While support for e-Governance and ICT started with military leadership in Korea, it has been sustained by the democratically elected leadership for the past 30 years. The gains achieved by each administration were not reversed by successive administrations, ensuring value to citizens. Aspiring countries should maintain their support for the e-Governance agenda over the long run, irrespective of changes in political leadership.

2. **Bureaucracy: Moving to a modern, innovative, technical and specialized public sector, from a traditional status-oriented model of career generalists, is essential to the design, development and management of complex ICT-based reforms.** As is the case in a number of countries in East Asia, the Korean public administration has its roots in a centralized and generalized Confucian bureaucracy. However, Korea made the transition from a bureaucracy of generalists to highly specialized, technical staff who could design, develop and manage technology programs for the public sector. Aspiring countries should recruit appropriately skilled staff and/or build technical capacity and skills within the civil service to steer the e-Governance agenda.

3. **Organization: Being willing to repeatedly reorganize and utilize presidential level inter-agency mechanisms helps to address the problems of horizontal and vertical coordination.** In many countries, government agencies tend to work in silos, making coordination across agencies a challenge. Korea tackled the challenge by readily reorganizing ministries and agencies and shifting responsibility for the e-Governance portfolio from IT-focused agencies to non-traditional agencies, such as the public administration and home affairs agency. However, these measures per se were insufficient to break down government silos. One successful strategy was the establishment of a supra-ministerial committee directly under the President to resolve inter-agency conflict and set a whole-of-government vision, priorities and tasks. Aspiring countries should experiment with setting up either a supra-ministerial committee or a team at the highest level of government to help steer the national e-Governance agenda and accommodate the constant shift in processes, accountabilities and demands, without being caught up in the politics of a particular ministry.

4. **Performance and Accountability: Making the transition from a wholly centralized governance model to local self-governance helps to bring public officials closer to citizen needs and concerns.** Prior to 1995, the heads of local governments were appointed by the central government and had little incentive to be accountable to citizens. In 1995, Korea made the transition to local self-governance, which brought elected officials and civil servants closer to citizens and enabled a greater capacity for innovating front-line citizen-service. One of the local e-Governance innovations, called G4C, which was developed for a specific district, was recognized and scaled up

box continues next page

Box 6.1 "The Korean Seven" *(continued)*

as a national solution to serve citizens across the country. Aspiring countries should provide space for local governments and frontline service delivery functions to innovate, serve and improve the quality of service delivery to citizens. These innovations may, in turn, serve as an invaluable source of inspiration for national programs that benefit all citizens.

5. **Architecture: Foundational technology components must be carefully sequenced and common government technology standards established.** Korea embarked on an e-Governance program in 2002, but was aware that the infrastructure building blocks were not equipped to facilitate the designs that were envisioned. In fact, the potential of e-Government was realized only after the Korea Information Infrastructure (KII), a high-speed broadband network, was completed in 2005, coupled with the development of a government-wide enterprise architecture (EA), an open source framework for all government agencies (eGovFrame), a public key infrastructure (PKI) for security, and an integrated government data center. Aspiring countries should carefully sequence the building blocks of core ICT infrastructure and ensure standardization across government agencies, when planning for a digital government program.

6. **Integration: Processes of government must be redesigned for citizens, and integrated across subnational and peer-level agencies.** Despite best efforts, countries around the world experience issues with integration of systems, as individual agencies find it more convenient to build systems in silos for specific needs, usually of the agency rather than the citizen or the beneficiary. Korea's strategy to deal with horizontal integration across peer-level agencies, vertical integration of agencies with their subnational counterparts, or citizen-centered integration, was to build an EA for whole-of-government and a common government data center, both mandated by law, to encourage agencies to share data and infrastructure across the board. This strategy enabled Korea to reduce the cost of investment in duplicated systems and to find efficiencies for government as a whole. Aspiring countries should find appropriate strategies to integrate data and infrastructure across government to solve the problem of inefficiencies caused, in turn, by issues of coordination across peer-level and subnational agencies.

7. **Implementation: Government needs to create new structures, such as public–private partnerships, to research, design, develop and implement projects for a digital governance program.** In many countries, the government assumes that the private sector has greater ICT capacity and thus, there is little focus on building that capacity or leveraging it for improvements in the public sector. The Korean government, however, made a conscious decision at the very beginning to work closely with the private sector in working groups and teams, building technology such as the electronic switching device, a domestic computer, the high-speed broadband network, EA, and an open source policy framework. Ultimately, the government was an investor in technologies that were developed and produced by private sector players. The Korean digital

box continues next page

governance approach not only created capacity within government, but it also helped nurture and create capacity within the domestic ICT industry. Aspiring countries should consider working closely with the private sector to develop their digital governance strategies, policies, frameworks and technology, as well as arranging staff exchanges as capacity grows.

Applying Korean Lessons to the Global Context

Korea has been extremely successful in transitioning from fragmented and/or outdated information systems to modern integrated digital solutions. It has also introduced non-technological reforms to complement and maximize the benefits of the technological developments. Along the way, it has had to deal with various challenges, thus providing depth to the lessons learned from its experience. Understanding these lessons will benefit countries in the early stages of implementing a digital governance program and also those in the process of planning how to move the e-Governance agenda to the next level.

We therefore apply these lessons to two groups of countries: **Group A: developing countries**, which require support for urgent needs (including countries with an extreme poverty rate above 40%); and **Group B: middle income countries**, which have more complex needs for transition to connected systems.

Countries in Group A typically operate within the first three stages of online service development (UN e-Government Survey 2014). They may be developing emerging information services such as information on public policy governance, laws, regulations and services, as well as links to ministry, department and agency websites. They may also be developing enhanced but limited digital services and downloadable forms to apply for administrative services and certificates. They may also have high levels of mobile penetration, compared to PCs, making mobile service delivery a viable option.

Countries in Group B are developing transactional services for two-way communication between the government and citizens, electronic identification for citizens, as well as financial (e-payments) and non-financial transactions (e-filing). In addition to transactional services between the government and citizens, more advanced Group B countries are developing connected services, which include Gov 2.0 and other interactive tools to request opinions and information from citizens, to integrate information, data and knowledge between government agencies in a seamless manner. They are also moving to a more citizen-centered approach, in which citizens are empowered to have voice and participate in decision-making (see figure 6.1).

Although all seven key lessons from the Korean experience are important for both Group A and B countries, we observe that some lessons are imperative for Group A countries while others are more relevant to Group B countries.

Figure 6.1 Stages and Demand for Support from Countries Implementing e-Government

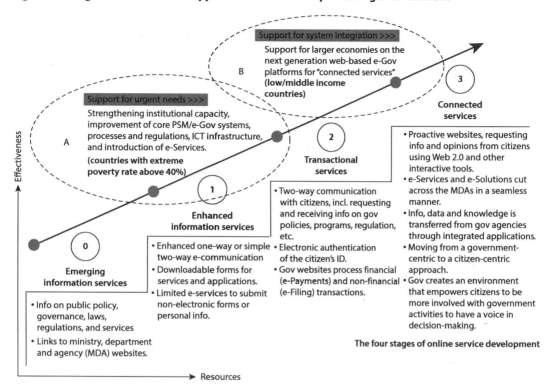

Source: The UN e-Government survey 2014; Cem Dener, World Bank.

Demand from Group A countries for advisory and technical support is typically for urgent needs such as strengthening institutional capacity, improving core public sector management and e-government systems, processes and regulations, implementing ICT infrastructure, and introducing digital services. Demand from Group B countries is typically for complex needs such as systems integration, including next generation web-based e-Government platforms for "connected services."

Key Lessons for Group A Countries

Key lessons from the Korean experience, for Group A countries relate to (a) the importance of political support for digital governance reforms, (b) the need to build hybrid technical/functional skills within the public sector, (c) the need to build innovative governance models to coordinate across agencies at both national and subnational levels, and (d) the importance of sequencing the development of core infrastructure components for a whole-of-government approach, including a common network infrastructure, EA, data center and open source frameworks.

1. Obtain Long-Term Political Support for Digital Governance

An overarching lesson for Group A countries is the need for sustained, high-level leadership and support for digital governance as a national priority defined by the government agenda. Korea is among a few countries that grew rapidly with relatively low income inequality. A uniquely Korean aspect of the story is that the push for a technology-driven development agenda was initiated by the military leadership, and subsequently nurtured over the long run by multi-party civilian leaders who recognized its value to furthering national and subnational governance outcomes.[5] Adult illiteracy declined from 78% in 1945 to 5% in 2008. Rural electrification rates went from less than 20% in 1964 to 100% in 1978. Tax administration reforms in 1966 marked a turning point in anti-corruption efforts and fostered a culture of integrity in civil servants, contributing to an increase in the share of domestic tax from 47% of government revenue in 1957 to 100% in 1974. Following on from these reforms were Informatization and e-Government plans, which were linked to the national development strategy (UN e-Government Survey 2014) and remained agnostic to the politics of successive administrations. Resolving the underlying constraints of low human capital and government failures with regard to economic development, including corruption, low tax collection rates and lack of clearly defined property rights, proved instrumental to the Korean miracle.[6] Over a 30-year period of national development and e-Government reforms, annual per capita income rose to US$27,000 (see figure 6.2).

Successive presidents of the Republic of Korea, from the late 1980s onward, emphasized the importance of ICT and e-Government, setting a clear direction

Figure 6.2 e-Governance and Economic Progress in Korea

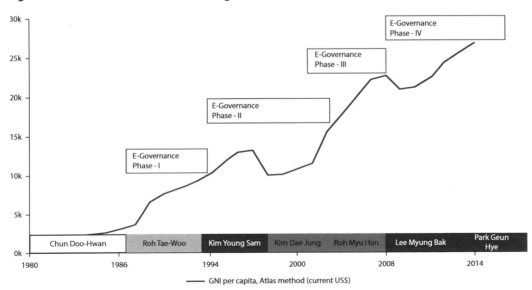

Source: Adapted from Moon 2014.

for public sector transformation. The motivation was to build an information-oriented e-society, with a view to enhancing national competitiveness and sustaining economic progress. The government provided funding and backing for technological advancement, giving impetus to a series of master plans drawn up by technocrats and enacting a number of laws and regulations.

Critical Success Factor: Sustained high-level leadership support across many terms of government, irrespective of political affiliation or changes in political governance

One of the concerns for the implementation of e-Governance reforms is that of continuing support from one political administration to the next, such that gains made by one administration are not reversed by the next. While the Korean model of sustained, politically neutral support for Digital Governance initiatives is instructive in this regard, a similar benchmark can be seen in countries such as the US and UK over a 30 to 40-year period. India, which has transitioned to a middle income country, has supported its digital governance agenda since the 1990s in the face of changes in political administration at both national and subnational levels. Nevertheless, there is a significant subnational variation in the quality of services and efforts to reform public services in India, often due to mixed electoral incentives and the threat to rents (Bussell 2012).

Policy Implications for Group A Countries: Tie the digital governance reforms to national development plans, and ensure support from legislation

The UN e-Government survey of countries suggests that e-Government reforms derive more value from Digital Governance programs when those programs are tied to a national development strategy and multi-year priorities. Furthermore, to ensure political continuity for reforms, it is essential to enact legislation and devise regulations to mandate certain actions. Laws and regulations need to address operational issues such as resistance to change and adoption of new IT systems/standards by agencies, and to create adequate infrastructure for e-Government, including high-speed broadband capacity (wired and wireless) throughout the country, especially in rural and remote areas. Laws and regulations will also need to address organizational issues such as developing an EA, establishing institutions to develop and manage the EA, and facilitating the creation of digital services and other aspects of the infrastructure for e-Government, including those related to mobile services, security, digital signature and privacy.

2. Build Hybrid Skills within the Public Sector
A second important lesson for Group A countries is the need for staff with hybrid (i.e., both technical and functional/sectoral) skills, who can contribute to the success of digital government initiatives.

In countries around the world, various assessments of public sector investment projects with ICT components indicate that a number of projects have failed, often involving significant time and budget overruns. This failure was not so much the result of flawed technology or weak design than the fact that these projects were managed by staff and executives who were not adequately equipped to handle complex projects with technological as well as non-technological content. In the absence of technical know-how, decisions tend to be political, leading to failed projects, messy contracting and the waste of public resources. According to Heeks (2006), "e-Government 'hybrids' steer a middle way between idolizing technology so much that it is the central focus of public sector change, and ignoring the technology so that it is unable to make a contribution to change." IT staff (who develop e-Government applications) and public officials, senior managers and politicians (who own or use the applications) are often in divided roles, creating a gaping hole between what is designed and what is delivered. To mitigate these risks, Heeks (2006) and others recommend the creation of hybrid staff, who combine information system competencies and public sector competencies to address the knowledge and skills gap between the two extremes of being a pure IT professional or a pure public sector professional.

Korea's approach in this context is noteworthy. With deep roots in a centralized model of public administration, Korea started out with generalists in the public sector. However, the government realized that advanced technical specialists in the public sector would be needed if digital government were to succeed; and over the years, public sector staff have shown a propensity to specialize, acquiring advanced degrees in highly technical subjects through overseas educational opportunities (see box 6.2). At the same time, there has been a steady flow and exchange of advanced IT skills from the private sector to the public sector (see chapters 2 and 3 on inputs to Korea's digital governance programs).

Box 6.2 Metamorphosis of the Korean Public Sector

The Korean public sector is modeled on centralization and the Confucian value of the pursuit of societal harmony over individualism, dating back to the Yi Dynasty (1392–1910). Confucianism, with its emphasis on moral rule and formation of an organization of generalists, led to a highly centralized authoritarian form of government under the ruling elite, in the interest of national identity (No and Ro 1993). State power is centralized under the leadership of the President, an approach that has continued under democratic rule. The development of a merit-based, technically-adept modern Korean state began over 50 years ago. In the early years, Korea had a "status-oriented" bureaucracy. Over time, the Korean public sector transformed into "modern" patterns of governmental organization (No and Ro 1993), showing a greater tendency to innovate, specialize, and differentiate, overcoming the culture of general public administration in order to become more technical and specialized. The gradual metamorphosis of Korean public sector staff from a career generalist to a more specialized technical profile over time is captured below in figure 6.3.

Figure 6.3 Korean Public Sector metamorphosis over the course of development

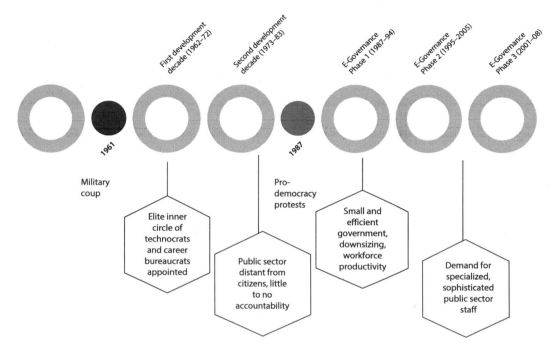

A key example of the public sector benefiting from this hybrid approach is the Tax Information Systems project, where IT experts and tax officials lacked mutual understanding of the other's knowledge and skills. Tax officials frequently and persistently resisted informatization and business process reengineering of tax processes, arguing that the manual methods were indispensable, while IT officials were constantly underscoring the importance of automation and coordination. To address these challenges, IT experts were deployed within tax departments and IT training was offered to tax officials. IT staff—computer experts, technicians and administrators—were assigned to the head, regional and district offices (Korea Eximbank 2013).

To secure the needed technical capacity to implement e-Governance programs, the government pursued three critical strategies.

1. *Training and education for public officials.* The government rolled out an extensive training and education program to develop a set of public officials and staff who embodied "hybrid skills and experience," including both sectoral/functional expertise as well as the ICT skills required to design and implement a complex whole-of-government ICT strategy.

2. *Digital literacy for citizens.* The government also invested heavily in digital literacy to promote the uptake of services provided by the government.[7] Campaigns and education programs were launched to bridge the digital divide for Korean citizens. In addition, universities and corporations

implemented IT learning courses and competitions. Ten million Koreans were educated in information technology through the government-led informatization education/training program from the late 1990s to early 2000s (see chapters 2 and 3).

3. *Collaboration with the private sector.* Finally, the government enhanced its technical capacity through the formation of "hybrid structures" such as public–private working groups and teams, which collaborated on the technology design of key ICT infrastructure, such as the electronic switching device, a domestic computer, the high-speed broadband network, EA, and an open source policy framework. The collaboration resulted in stronger government capacity to implement e-Government and stronger private sector capacity, evidenced by the numerous Korean companies that are now global players in the ICT industry.

Reflecting on Korea's approach to building human resource capacity and skills, it is evident that they focused equally on the upstream impact of e-governance on private sector suppliers of services and on the downstream impact of e-governance on the public sector staff and its clients, i.e., citizens. This is an important lesson, particularly for Group A countries.

Critical Success Factor: Cultivation of technical capacity within government

Governments need to have skilled staff for their digital needs. However, governments around the world, from middle income countries such as Indonesia to low income countries such as Tajikistan, are concerned that the public sector skill-set has failed to keep pace with technological changes, notwithstanding uncompetitive salary structures to attract such skills. These governments are attempting to draw cutting-edge private sector skills and staff to the public sector through partnerships and special staffing programs. By creating hybrid staff profiles, Korea created teams who could work together because they combine on one side, sector specialists with enough ICT competencies to talk to ICT specialists, and on the other side, ICT specialists with enough sectoral knowledge to talk to sector specialists. A similar benchmark can also be seen in countries such as the US and UK, which are demonstrating the importance of leveraging technology and management skills into the public sector. In the US (18F under the US General Services Administration and the US Digital Services, a White House agency to coordinate across government to modernize IT systems) and in the UK (a Government Digital Service), renewed functional-technical teams are leading digital transformation initiatives for the government. These examples should serve as an impetus for aspiring governments to invest in people with technical skills to build careers in government.

Policy Implication for Group A Countries: Create a "hybrid" set of knowledge and skills in the government, and provide digital literacy programs for citizens

The public sector is in dire need of designers, developers, and digital strategists who can design and build with an intimate understanding of the organization, as well as knowledge of public procurement and contracting, while creating institutional knowledge and sustainability. It is an investment in a modern and agile institution with citizen-focused products and services. This new combination of skills includes those that are sectoral/functional in nature and those that are IT-related. Group A countries that are laying the infrastructure foundation should employ strategies that involve training and educating generalists and functional specialists in IT knowledge and skills, and, when necessary, recruiting new "hybrid" staff. Group A countries that are starting to develop basic digital services should aim to train, educate and acquire staff with skills and experience. They should ensure that relevant officials and staff with "hybrid" skills have specialized technical expertise. This requires knowledge and skills involving automation of both front- and back-end service delivery. Furthermore, these programs should be complemented with digital literacy programs for clients and consumers of public services, i.e., citizens.

3. *Experiment with Innovative Governance Models to Coordinate across Agencies*

A third key lesson for Group A countries is the need to improve inter-agency collaboration in building a "Virtual State," which is defined by Fountain (2001b) as "a government in which decision makers increasingly use information technology (IT) in ways that blur the boundaries among agencies, levels of government, and the private and nonprofit sectors." Subsequently, Fountain (2009) observed that agencies did not employ IT to rewrite the rules of Weberian administration but, instead, used IT to build power by shoring up technology and people, ultimately reinforcing silos. Contrary to business perceptions that IT always was associated with resource and cost reduction (i.e., evidenced by a "virtual" Amazon outcompeting a "bricks & mortar" Barnes & Noble), IT innovation in the public sector did not eliminate intermediaries in government, cut out organizational structures, or downsize staff. Instead, IT created new intermediaries, with the intention of ensuring the success of the program and supporting collaboration. Ultimately, e-Government did not break down silos, but resulted in the building of more layers and roles in government.[8] This is supported by literature reviews and meta-analysis, analyzing the causal connection between IT and organizational change. They suggest that, although the goal of e-Government programs was to improve administrative efficiencies and quality of services delivered to citizens, the benefits of IT did not substantially alter the underlying form of administrative organization, practices or behavior.

In Korea, agency coordination challenges were addressed through a whole-of-government approach, harmonization of data and definitions across agencies, and establishment of budgets for expenditure on IT. An influential governance structure at the supra-ministerial level led by the President or the Prime Minister's Office was established in response to the challenge of coordination and collaboration. This helped to virtually join up agencies and resolve inter-agency conflicts.

Nonetheless, coordination across public agencies in Korea has continued to be a challenge throughout the Digital Governance journey, in part due to competing agency interests. Overcoming these parochial interests has not been easy, and remains a challenge today. The government's response was to apply a trial-and-error approach to managing these tensions and resistance. One organizational approach would be tried; if it was not working, the government would quickly shift to another. In fact, over the course of a few decades, many organizations were established and charged with managing all or part of the e-Government program. No one entity or organizational structure provided "the solution." Furthermore, what worked at one point might be different than what would work at another time. What is most important is to recognize that Korea continued to try one structure after another until it found something that worked. This flexible and tenacious mind-set played an important part in managing the organizational challenges of the e-Government program (see box 6.3).

Korea's commitment to anticipate, plan for, and address inter-agency conflicts has been a critical component of its management strategy and leadership approach. This commitment, combined with the government's flexibility, tenacity and emphasis on performance and results, has played an important

---

Box 6.3 The Organizational History of e-Government in Korea

One of Korea's key responsible agencies for e-Government, the Ministry of Information and Communication (MIC), operated as an integrated ministry between 1995 and 2007. However, conflicts regarding e-Government and the ICT industry arose between MIC and other ministries, such as the Ministry of Government Administration and Home Affairs (MOGAHA)[9] and the Commerce, Industry and Energy Ministry. For instance, one of the conflicts related to the jurisdiction of the government computing center; this conflict lasted several years until the direct intervention of the President. By 2008, the functions of MIC had been redistributed to mitigate conflicts between MIC and other Ministries on e-Governance and ICT policy. Implementation of e-Governance was moved to the Ministry of Public Administration and Security (MOPAS)—previously MOGAHA, and the National Information Society Agency (NIA) was moved under MOPAS (see figure 6.4). When MOPAS took on the lead role as implementing agency for e-Governance, a Chief Information Office Council (CIOC) was established, comprising Assistant Ministers from 25 Ministries and Agencies, and chaired by the Minister of MOPAS. The role of the CIOC was to formulate and implement e-Government policies, share administrative information, establish a government-wide EA, and standardize ICT across all agencies. In theory, this was an effective method to achieve a whole-of-government approach. In practice, however, the CIOC failed to achieve these goals because of the lack of specialized technical experts from ministries and the existence of higher priority issues for representatives on the council. In contrast, the Presidential committees at the supra-ministerial level, chaired by the Chief of Staff to the President or the Senior Secretary for Policy and Planning, proved to be more empowered and effective in resolving inter-ministerial conflicts and building a cooperative environment. Technical staff from individual ministries would often refer their issues directly to this body, based on concerns of policy neutrality with MOPAS as a peer-level agency.

Figure 6.4 Restructuring for E-Government in 2009

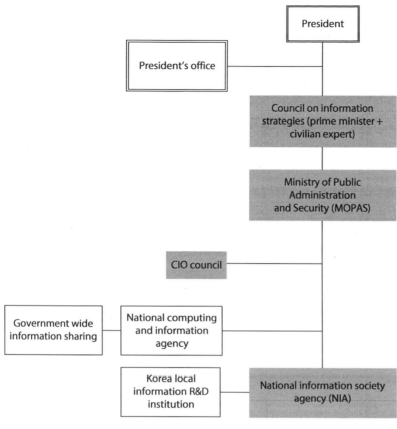

Source: Song and Oh. 2012.

role in overcoming agency resistance-to-change, which is a fundamental dynamic of any e-Government program. Korea pursued two remedies for addressing inter-agency conflict. First, the government created a high-profile organizational venue to tackle conflicts between agencies. Second, agency-specific systems were reconciled through vertical (subnational) and horizontal (peer-level agency) integration of the EA, an infrastructure shift that forced agencies to work together. Today, EA is almost fully integrated across government; and, while agency tensions continue, they are mitigated and managed through the oversight of a supra-ministerial committee empowered to resolve conflicts and to establish policy positions through the authority of the President or the Prime Minister, given the national priority of the Digital Governance agenda (see box 6.4).

An example of how **national peer-level agency coordination** methods worked within the context of an e-Governance project's implementation is illustrated by the Social Security Information System (SSIS). When this system was being built, disagreements arose between the Ministry of Interior (MOI) and the Ministry for Health and Welfare Services (MHWS). The disagreements related to whether the system would be managed at the central government level by

---

Box 6.4 Governance Models for Horizontal Coordination (across Peer-level Agencies)

The Korean government experimented with a number of different governance models to coordinate horizontally across agencies. In the first stage, when the National Basic Information System (NBIS) project was launched, it was managed by an Information Network Supervisory Commission (INSC) chaired by the Chief of Staff to the President. The Chief of Staff played a key role, representing the President, influencing and coordinating across agencies and pushing forward implementation of the project. The role was critical to resolving inter-ministerial conflicts, supervising measures for common standards across government, ensuring security, sharing ICT resources, and securing financial resources (Korea Eximbank 2013).

In the second stage, during the development of the KII and high-speed broadband infrastructure network, an Informatization Promotion Committee (IPC) was set up, chaired by the Prime Minister and comprising 24 Ministerial-level representatives from all related agencies. The purpose was to review and coordinate informatization efforts based on the National Informatization Framework Promotion (NIFP) Act (Moon 2014). MIC was the lead agency to marshal financial resources—the Informatization Promotion Fund (IPF)—as well as technical and human resources. Funding for the IPF came from the private sector, namely telecom operator profits. In terms of policy coordination, a high-level President's Council on National Information and Communications Technology led the development of a strategy and evaluation of initiatives. The council was chaired by the Prime Minister and consulted extensively with an Advisory Committee, comprising experts from industry, academia and research institutions. Liaison officers (the Chief of Staff to the President or the Senior Secretary for Policy & Planning) were appointed to facilitate communication between committee members and policy makers (Song and Oh 2011: 25–30) (Korea Eximbank 2013).

In the third stage, during the development of e-Government systems, a presidential Special Committee for e-Government (SCeG) was set up, chaired by the Senior Secretary to the President for Policy & Planning. The SCeG consulted extensively with external technical experts and the private sector, and brought together MIC, the Ministry of Government Administration & Home Affairs (MOGAHA), and the Ministry of Planning and Budget (MPB). The SCeG was replaced by the President's Council on National ICT Strategies following the enactment of the NIFP. It was co-chaired by the Prime Minister and a civilian expert. The MOPAS replaced MIC as the lead agency (Song and Oh 2011: 36–37).

MHWS or through local governments using a pre-existing system called the SAEOL Administration System developed by MOI. Other agencies, such as the Supreme Court, Korean Tax Services, Korea Pension Service, National Health Insurance Service, and the Korean Employment Information Service, were concerned about sharing with SSIS information related to tax, income, property, and travel held in their respective systems. These tensions were mitigated by the Prime Minister's Office, which convened meetings with the agencies concerned; established policy positions on the management of the system, making MHWS the responsible agency; created a division of labor between different levels of local government at the City, County, District and Sub-levels; incorporated relevant aspects of the existing SAEOL Administration System; and integrated more than 50 different types of public information from relevant agencies (219 kinds of social service related information managed by 27 public agencies) into the SSIS (Korea Eximbank 2013).

To address the challenge of developing an integrated approach across public sector agencies, financial coordination methods were also employed (see box 6.5).

There were challenges with regard to **vertical coordination between the local and national government** at the inter-governmental, inter-agency and inter-sectoral level (Kim, Lee, and Kim 2009). Although the South Korean Self Governance Act decentralizes administration to local governments, these e-government initiatives required approval from higher levels of government to resolve associated legal and institutional issues. The study of Gangnam-gu's e-Government initiatives by

---

Box 6.5 Innovation Financing, Budget Allocation and Prioritization

Financing is one of the important issues to be considered in order to efficiently develop a Digital Governance program. In Korea, initial funds were supplied through the IPF, which tapped into both public and private sources of financing and where profits were ploughed back into the ICT sector (Moon 2014). A considerable amount of financing was raised (albeit reluctantly) from national telecommunications operators, where the government had invested significantly in R&D, and which had been privatized.

The IPF, established in 1993 under the IPC, was an innovation that allowed MIC to coordinate the policy agendas of different public institutions (Song and Oh 2012). The IPF was a special purpose fund that provided operational flexibility for investments in ICT in Korea under the budgetary constraints of single year, line-item budgeting. It allowed ministries and agencies to spread the risk of investing in ICT projects by providing 50% matching finance for projects, and to draw on technical expertise from the NIA to manage and implement projects. Initially the IPF financed national database projects (1987–1992), followed by informatization projects (1993–2004), and then R&D (2005–present).

The IPF also served as an IT Governance mechanism to prioritize investments and spending across all agencies and to consider projected streams of expected returns. It provided a longer term window of financing for projects to ensure sustainability. In addition to the large projects, the IPF considered innovation projects and key demonstration projects, providing space for creativity within the public sector. To spur government agencies to collaborate with each other, performance incentives were instituted to deliver projects by providing flexible and matching financing from the IPF, through a medium- to long-term budget plan of at least five years. The fund tapped into both public and private sources of financing and profits were then ploughed back into the ICT sector (Moon 2014). Later on, during the e-Government maturity period, the IPF was abolished and an Information and Communication Promotion Fund was established. Financing of e-Government projects was carried out by surveying the demand for e-government activities from individual ministries, followed by deliberations with related agencies including the Ministry of Planning and Budget.

Bretschneider et al. (2005) found a lack of collaboration among different levels of government; it also found that weak coordination amongst local divisions was a barrier to implementation. These challenges remain pertinent today.

In summary, it is important to note that central institutions in charge of implementing e-Government were restructured frequently, while responsibility for the portfolio shifted from MIC to MOI. Such changes were not based on whim. Political and senior administrative leadership took into account feedback from frontline implementation and collaborating agencies regarding inter-ministerial conflicts. These issues were resolved through the Chair of a high-level supra-ministerial committee—either the Chief of Staff of the Prime Minister's Office or the Senior Secretary to the President for Policy and Planning—who would mobilize agencies and coordinate across government, with the strong will and support of the President for e-Governance as a national priority.

Critical Success Factor: Establishment of a body at the highest level of government to manage cross-agency coordination

Clear lines of responsibility and accountability should be established for e-Government design and execution at the agency level. In Korea, the high-level coordination body provided leadership and oversight of government-wide automation of services. This ensured that a "digital by default" strategy was used as one of the primary methods for innovating, enhancing access for citizens, and improving the quality of service delivery, across all agencies. Lack of cross-agency collaboration is a common concern, but Korea overcame this by leveraging the high-level coordination body to resolve issues and conflicts and successfully implement digital services.

Policy Implications for Group A Countries: Adopt an experimental, iterative, learning approach that enables the development of an appropriate structure to facilitate prompt implementation of solutions throughout the public sector

A common issue that many governments face is choosing an appropriate governance model to manage e-Governance reforms. Should the governance structure be a permanent one in the Prime Minister's Office, as for example, in the UK, or should it be housed in the ICT Ministry, such as in India, or in General Services Administration, such as in the US, or in the Ministry of Finance and the Prime Minister's Office, such as in Australia? Many governments suffer from analysis-paralysis—spending too long on working out the correct path prior to implementation, and then getting stuck with one model or the other. However, governance structures are evolutionary and they are expected to evolve and shift as needs change. Korea achieved success through adopting an experimental, learning approach, which entailed modifying the initial path if it did not deliver the anticipated solutions. Aspiring countries should develop multiple iterations of a governance model to arrive at an appropriate structure to oversee the national digital governance program. Importantly, the structure should help arbitrate and resolve inter-agency conflict, set a whole-of-government vision, and successfully steer the agenda without being caught up in the politics of one ministry or the other.

4. Sequence the Development of Infrastructure and Services

The fourth key lesson for Group A countries is the importance of sequencing the development of key infrastructure components, and, at the same time, establishing common standards for government agencies building e-Government systems. While e-Government projects were originally articulated in the early 2000s, the development was not linear. Korea quickly realized that the development of a high-speed broadband network would be critical to the program's success. The development of network infrastructure paved the way for the implementation of e-Government projects by the mid-2000s (see box 6.6).

Box 6.6 Korea's Sequencing of Technology Layers

The groundwork for Korea's network infrastructure—the NBIS—was put in place almost 10 years prior to the development of its broadband network—the KII. In 1986, Korea established the NIA to develop the standards for the implementation and operation of the NBIS and, subsequently, to supervise the project. The vision was for the NBIS to consist of five digital networks to facilitate communication for government and government-funded institutions; banks, insurance companies, and the securities commission; universities and research institutes; defense-related organizations; and security-related organizations. Of the five digital networks, it was the administrative or government network that formed the core of the NBIS project. The implementation of KII, built between 1995 and 2005, gave fillip to the ambitious e-Government program envisioned in the early 2000s. With the launch of the "Giga Internet Service Plan" in 2009, a plan to establish a Ubiquitous Sensor Network (USN) that can link services with Radio Frequency Identification Devices (RFID) and sensors that form the mesh of the "Internet of Things," Korea prepared to deliver high-speed, high bandwidth services over mobile devices.

While the **network layer** was being built, Korea turned its focus to building up a national **database layer**. Paper records for six types of government-related work, namely resident registration management, real estate management, employment management, customs clearance management, economic statistics, and automobile registration management, were digitized into databases for online public service delivery. The policy focus was then turned to creating a **process layer** by connecting, integrating and reengineering business processes[10] that were previously disconnected and duplicated across ministries and agencies. These initiatives were supported by the Electronic Government Law of 2001, which stipulated the improvement of citizen convenience, administrative information sharing and business process innovation to improve the efficiency and productivity of public administration. 11 major e-Governance projects and 31 priorities were launched between 2001 and 2007. The purpose of these initiatives was to develop **integrated applications**: (1) to simplify and automate document management within government; (2) to improve and interconnect

box continues next page

Box 6.6 Korea's Sequencing of Technology Layers *(continued)*

government financial information through the Digital Budget and Accounting System (DBAS)[11] (3) to manage local government information (SAEOL); (4) to manage human resource information; (5) to manage education information (NEIS); (6) to construct integrated government-wide data centers; (7) to develop a citizen-services one-stop window and government portal (Minwon 24); (8) to allow citizens to submit national taxes via Internet (HTS); (9) to integrate and automate government procurement (KONEPS); (10) to develop an interconnected system for four social insurance programs—pension, health, accident, and unemployment; and (11) to develop an e-authentication system.

Reflecting on Korea's approach, one key factor that contributed to the success of the e-Government program was getting the sequencing of technology layers right. At an elementary level, it appears that a network layer was first built, followed by a database layer, a process layer and then an integrated applications layer, after which, government standards were developed through an EA approach, an open-source framework and a common data infrastructure (see annex 6B).

In retrospect, the sequencing of technology investments appears not to have been so linear. When Korea launched its informatization program in the late 1980s, there was almost no demand and there were very few willing investors for a nationwide high-speed broadband network. It was a classic chicken or egg dilemma—could services be developed without network infrastructure, or could investments in network infrastructure be made without demand for services? With a flexibility that has come to characterize their approach, the government of Korea laid the groundwork for broadband network investments, and articulated a number of e-government priorities and projects to deliver services to citizens, businesses and back-offices. Strategically, government institutions became one of the largest sources of demand for applications and services, thereby stimulating private investment in a broadband network to support the delivery. By the mid-2000s, once massive high-speed network infrastructure was made available, there was an explosion of online applications for e-Government, banking, gaming etc. The proliferation of applications and services in turn prompted the quest for common standards and a whole-of-government approach to address issues of coordination, duplication and redundant investments.

Given that correct sequencing is a critical factor in ensuring the success of an e-Governance program, we examine this issue in more detail in box 6.7 and annex 6A. Box 6.7 considers Korea's experience, while annex 6A aims to provide countries at different levels of development with guidance related to the sequencing of an e-Government initiative. This guidance assumes that implementation occurs over decades, requiring top-level leadership of government, regardless of political orientation. The chart has a straightforward framework depicting three stages of development: (a) laying the foundation; (b) introducing digital services; and (c) building for whole of government. For each of these stages of development, the framework discusses the governance and technology

---

Box 6.7 The Importance of Sequencing

The Korean experience indicates that sequencing is a critical factor in building a successful e-Governance program, and provides insights into the appropriate sequencing of e-Government development and investment over decades and through multiple political administrations, a sequencing centered on the use of e-Government to further national development goals.

Over five decades, Korea developed ICT as a means of supporting its national development goals, collaborating closely with the private sector. The system was rooted in a strong legal framework, and it was bolstered by effective organizational, human resources, and local government capacity. Further, with clarity of vision and commitment, Korea developed an EA which allowed for standardization and cross-government integration. Yet these components did not fall into place randomly. A critical component of Korea's success has been its ability to get the sequencing right, through an iterative and flexible process.

Today, this issue of sequencing remains a practical challenge for virtually all countries developing e-Government plans, organizations, systems, and architecture. With proper sequencing, development opportunities expand exponentially; with poor sequencing, costs are high, yet yielding negligible results. Given the complexity associated with technological innovation, the cost of ICT investment, the rapidly changing nature of technology, and the opportunity to leverage ICT to improve the livelihood and well-being of citizens around the world, sequencing becomes a critical issue for ensuring that ICT investment achieves development results. With this in mind, we draw on the Korean experience to shed light upon sequencing. While e-Governance involves extensive experimentation and learning-by-doing, there is a natural progression of investment and activity associated with the creation of a holistic e-Government system.

---

strategies to be pursued, including those related to establishing a legal and regulatory framework; developing the appropriate human resources capacity to develop and implement a digital governance effort; organizing the bureaucracy to effectively plan, coordinate and manage an e-Governance initiative across government; empowering local government with citizen-centered solutions; and creating the architecture and infrastructure required for digital governance, including the strategies required to develop an integrated system and the implementation approaches to leverage private sector capacity.

Critical Success Factor: Appropriate sequencing of citizen service innovations to help stimulate demand for the core technology infrastructure needed to sustain reforms

Korea's experience shows that the development of e-Governance programs is not always linear. There is a tension between "build (infrastructure) and they will come" and the crafting of demand from citizens for digital service delivery. Depending on the context, there have to be tactical, effective ways of stimulating demand for services. Take for example, Bangladesh, which, in spite of deep infrastructure challenges, has targeted the electronic delivery of services to the poor through citizen service centers and digital services at the local level through an A2I program. Bangladesh set up 4,547 digital centers, providing 200 million services[12] by nurturing innovation and simplification of service delivery processes. Meanwhile, national initiatives to bring broadband connectivity and other core technology elements are in still in the early stages of financing and capability to support these much-needed service delivery goals. Bangladesh sequenced its service delivery program, targeting grassroots and bottom-up innovations to benefit frontline government officials and citizens, which they expect will in turn stimulate demand for the infrastructure and core foundational technology that needs to be built for these reforms to succeed and be sustained over the long term. In Korea's case, success lay in its capacity for experimentation, and the ability to leverage private sector investments to solve the chicken or egg dilemma of sequencing the ICT infrastructure underpinning service delivery reforms.

Policy Implications for Group A Countries: Korea's experience with sequencing technology infrastructure and services offers a number of important policy implications:

- *Technology Roadmap and Approach:* Develop a Technology Roadmap and Approach that clearly sets out the key infrastructure and service building blocks, and carefully selects and sequences the components that satisfy demand, outlining when and how they will be built. These elements are critical for scaling up across government. The creation of adequate high-speed broadband capacity (wired or wireless) across the country to serve government agencies (public), citizens and businesses (private), and research centers/academia is foundational.
- *Front- and Back-end Development for Streamlined, Citizen-centered Service Delivery:* Develop front- and back-end systems to support a citizen-centered service delivery system, focusing on reducing citizen time, cost and number of visits required to secure a service. Integrate citizen feedback and grievance redress mechanisms for each service into the system design. Track service delivery progress and capture data for delivery improvement for each service. Leverage mobile devices for data collection, notifications and tracking, citizen feedback, and for exchanging photographs and videos if applicable.
- *Citizen-centered Process Redesign:* Redesign processes for the convenience of citizens, and include citizens in the design process (digital by design), so that citizens will not have to interface with multiple agencies in order to receive information or services or to interact with the government. The goal is for government agencies to work together seamlessly to deliver services to citizens through the most cost-effective channels—online, government office counter, call centers, other organizations (outsourced co-providers) and so forth.

Key Lessons for Group B Countries

Group B countries at low and middle income levels are typically at various stages of advancement on the foundational issues addressed in the previous section. Such countries are generally looking for advice on enhancing their digital services transaction capacity and becoming a more data-driven and service-oriented government. Three key lessons for Group B countries, from the Korea experience, are: (a) the importance of empowering local governments to develop a more citizen centered and service-oriented government; (b) the need to integrate systems across the national and subnational levels for a whole-of-government approach; and (c) the need to utilize public–private partnerships to inject innovation and competitiveness into the implementation, and achieve better performance outcomes.

1. Localize Accountability and Build a Service-Oriented Culture for Citizen Service Delivery

Governments in Group B countries are typically developing digital services for their citizens, and the more advanced ones are developing connected services for a more data-driven and service-oriented government. A key lesson for this group of countries is the importance of a local governance model that increases front-line public officials' interaction with citizens and heightens their understanding of citizens' needs.

In Korea, the initial e-Government systems were proposed at the national level, and these were core institutional systems to improve administrative efficiency. The move to a local self-governance model provided an impetus to develop front-end e-Government systems that would directly benefit citizens. Notably, such systems were sustained by frontline public officials, who would actively look for ways to improve the quality of services delivered to citizens. It marked a departure from a culture of seniority based promotions to an innovation-focus and an orientation change to a benign service delivery[13] mentality (see box 6.8).

If we reflect on the innovations at the local government level that improved government performance, we see that employee motivation was one of the contributing factors to its success. Motivation can be negatively affected when computerization initiatives raise expectations of benefits to employees in the form of workload reduction, and yet the workload increases, partly due to the running of parallel systems—both manual and computerized (Bretschneider et al. 2005). For instance, the actual functioning of the local e-participation system reflected the concerns of officials, who would place online only those documents or meeting summaries that they deemed important; less important documentation was not included in the system, partly because it was time consuming to upload. The mayor actively utilized an incentive-based points system to motivate employees to improve administrative efficiency and quality of service delivery and to recognize commitment to innovation to serve citizens. The performance incentive system for employees provided cash rewards, overseas educational opportunities and promotions, based on the level of contribution to

Box 6.8 e-Government at the Local Level in Korea

Following massive pro-democracy demonstrations in 1987, the South Korean Self Governance Act on decentralization[a] was passed by the National Assembly. This Act provided an impetus for local governments to use e-Governance to improve responsiveness and accountability related to citizen needs. It ended the practice of centrally-appointed top bureaucrats serving as the head of local government, and shifted accountabilities to locally-elected officials, lessening the gap between citizens and their public service providers. This shift incentivized local government to develop and deliver citizen-centered solutions that met the needs of citizens and reflected their priorities and concerns.

Early e-Governance applications required citizens to go to district or local offices to receive government-issued documents related to resident registration, real estate and vehicle registration. Under the national e-Governance initiatives, institutional databases were established and networks were built to connect agencies. The central government provided resources to support local governments, resulting in the building of nationwide IT infrastructure to facilitate the adoption of e-Governance, and promote transparency and citizen participation at the local level. Democratic political leadership and public pressure provided an impetus to utilize e-government as a tool to establish a more service-oriented government (No and Ro 1993). The move to local self-governance prompted the design and development of services that would directly benefit citizens, improve the quality of public services, and make local government accountable. Two examples of local government initiatives are:

1. **The Smart Gangnam Cyber City Project** developed applications for civil registration, permits, real estate, payments and traffic fines. One of the applications pioneered in Gangnam-gu was a "G4C" system. Using a web-based portal, it provided citizens access to welfare-related insurance services, job training networks, tax services, and a range of government-issued documents. (Bretschneider et al. 2005). Given the success of the G4C citizen service, the innovation was scaled up at the national level in 2010 to improve service orientation and to create a more citizen-centric government. The nationalized system, named Minwon 24, covered 40 million Koreans, provided authenticated e-signatures to over 10 million citizens and 1 million public servants, and expanded the range of services and transactions that were accessible over the internet.

2. **The Seoul City Government's OPEN system** was designed to improve accountability. Linked to 870 government agencies and 260 local governments, OPEN is an online disclosure portal providing a wide range of information to address corruption. It provides administrative information, allowing citizens to keep an eye on the processing of permit applications, especially in the areas where irregularities are more likely to occur (Choi 2013b). While citizen use of the system is limited, the information and documents are transparent and publicly available, thus encouraging non-elected public officials to be "more cautious" about their activities and creating a sense of accountability (Bretschneider et al. 2005).

a. There are three levels of government in Korea: (a) the central government; (b) Seoul City, 6 Metropolitan Cities and 9 Provinces; and (c) 25 Districts under Seoul City; 44 Districts and 5 Counties under the Metropolitan Cities; and 71 Cities, 94 Counties, 188 Towns and 1191 Townships under the Provinces. Local legislative council elections were held in 1991 and elections for Mayors and provincial Governors were held in 1995 at the City (Si) level, and the district (Gu) and county (Gun) levels, respectively. There is a further sub-administrative level of districts, villages, towns and county subdivisions, but these leadership positions are not elected.

improve the quality of citizen services. The rationale for the incentives program was to shift the focus from a seniority based promotion system to that of an innovation-focused system where employees would actively look for ways to improve the quality of services delivered to citizens. A study of the program revealed that the performance incentive programs indeed motivated employees to put in greater effort in Digital Governance; however, it did not improve collaboration among divisions to deliver services successfully (Kim 2008). Nevertheless, the interventions enjoyed a greater sense of ownership and benefited from the improved performance of citizen-facing officials, offering autonomy (or the desire to be self-directed), mastery (the desire to keep improving at an important task), and a sense of purpose (the desire to find meaning beyond oneself) (Pink 2011).

Critical Success Factor: Employee Motivation, Innovation and Commitment to Deliver Better Quality and More Efficient Services to Citizens

Digital is increasingly the way citizens interact with government. From submitting passport applications online in Armenia, to paying parking tickets through mobile phones in Estonia, prior in-person interactions are now occurring online. A digital interface is only the first piece of a bigger service delivery process—one that occurs offline in government offices where transactions are processed, decisions are made, and services are distributed. The digital experience is the front door. It is what lies behind the door that really changes the lives of people and communities (Karippacheril and Tavoulareas 2014). However, seldom emphasized is the importance of motivating frontline service delivery providers to adopt a culture of innovation and service that makes the citizen the top priority. Aspiring countries should develop people, processes and systems that are optimized to understand citizen needs in their particular context, and should then ensure that these needs are addressed efficiently in a way that builds trust between the state and the citizen.

Policy Implications for Group B Countries: Establish local government as a key player in the provision of digital services, empower it to make citizen services "digital by default," and build a service-oriented bureaucratic culture

Local governments are well placed to use e-Government interventions to improve responsiveness and accountability related to citizen needs. Local governments should therefore be empowered to develop digital services by default to reduce citizen cost, time spent waiting for processing, and the number of visits they must make to government offices or to citizen service centers to secure a service. Group B countries that are developing "connected" services, including next generation Gov 2.0 and Gov 3.0 platforms, should emphasize the role of local governments in designing services with the help of citizens so it is more useful to them. Citizen engagement with local governments should be intensified by using data generated from transactions to support a more evidence-based

Bringing Government into the 21st Century · http://dx.doi.org/10.1596/978-1-4648-0881-4

system of improvements and performance management through citizen feed-back. Data should be integrated and interoperable across different local and regional governments to compare the performance of service delivery across government.

Building a service delivery culture, as a policy goal and goal of e-government, should be deeply emphasized for bureaucracy. Service orientation can be established through historical tradition, political leadership, or democratic pressure from citizens. Since the 1990s, the Confucian tradition of people-based government or government for the people in Korea, although not government by the people, as well as democratic political leadership and democratic pressure in society, has provided a social atmosphere that has supported the use of e-government as a tool for a more service-oriented government in Korea.[14]

2. Integrate by Building Common Standards for Whole-of-Government

A perpetual challenge that Group B countries struggle with is the smorgasbord of applications, services, business processes, databases, technology and other infrastructure that proliferates in the absence of coordination among government agencies. Korea was no different in this respect, with a centralized bureaucratic structure that made coordination with peer level and subnational agencies challenging. The e-Government initiatives of the early 2000s resulted in agencies building systems independently with different specifications and without concern for interoperability, compatibility, duplication, redundancy, or a whole-of-government approach. As a consequence, the government mandated an EA framework through law, to integrate business processes, minimize redundant software development, and share and reuse common applications, processes, data, hardware, software and security resources across government (see annex 6B).

Reflecting on Korea's approach to integration, a legally mandated EA framework for all government agencies became key to creating seamless services. It helped drive reforms of administrative procedures, public services and information resources. Korea utilized the EA approach as a strategic planning tool to create linkages between government agencies and to improve interoperability of processes across agencies, thereby improving public service delivery to citizens. Korea's use of a national level EA approach, mandated by law, helped with the management of public ICT resources. Not only did it create checks and balances on overlapping agency investments, but it also helped with interoperability between agencies. It allowed for easy adjustments, alterations and upgrades to be made when a system's function or requirement was changed, since the system would have been established in accordance with a design plan with a common reference or standard. This is referred to as an increase in interoperability. Interoperability increased even more since it was based on an open source framework. Korea adhered to an open source policy as the use of open source software allowed for the sharing of components, dramatically reducing the cost and time necessary for development and maintenance.

Critical Success Factor: Adoption of a National Level Enterprise Architecture Approach

Countries such as Moldova and Estonia have learned from experience that the lack of coordination and collaboration across agencies leads to poor integration of systems and limited exchange of data, creating silos within government. It also hampers the design and implementation of an end-to-end administrative process for delivering digital public services. In Moldova and Estonia, administrative procedures were often transferred from paper-based to electronic systems without rationalizing, simplifying and streamlining the underlying administrative processes, functions, and services. These weaknesses were compounded by institutional issues of horizontal coordination across public agencies. To address the problems, both Moldova and Estonia developed a government-wide EA framework, applying national common standards for interoperability to enable e-Governance-based reforms. In general, however, it must be said that, EA has been difficult for governments to develop or implement in practice. The goal is to draw up conceptually what system components will be developed and how it all fits together. EA is typically applied to e-government systems that are of a certain size, because the time and effort undertaken to develop an EA may be hard to justify for smaller sized systems. A considerable number of e-government projects in developing countries are small. In such cases, the decision to apply EA by aspiring countries should be made carefully.

Policy Implications for Group B Countries: Integrate national and local government systems, emphasizing interoperability, compatibility, and elimination of duplication and redundancies, so as to strengthen service delivery

- *System integration to improve service delivery.* Building a "service-oriented" government means that the integration of cross-agency systems and processes should ensure end-to-end service delivery to the citizen. Projects and priorities should be established to promote integration of back-end systems and elimination of manual and offline touchpoints, failing which visually appealing front-end systems will become window-dressing for citizens. Examples of system integration projects include automated document management, integrated financial management, local government information management, human resources information management, and integrated procurement management.
- *An architectural foundation to establish "Data-driven governance."* A culture of "Data-driven governance" should be established within government, which means that governments use evidence—i.e., data and analytics from transactions—to make policy decisions.[15] In order to do so, countries should establish an architectural foundation for intra- and inter-governmental connectivity and ICT resource management, as well as common standards for government agencies in building e-Government systems.

- *Standardization to minimize costs and leverage data sharing.* An emphasis on standardization is key in order to minimize costs associated with the integration of software, processes, and systems across agencies. Standards include an Open Source framework for developing the software required for e-Government systems, public key infrastructure for security, digital signatures to eliminate paper-based transactions, digital identity to authenticate citizens, and a data center to harmonize ICT resources, among others. The Open Source approach avoids dependence on private technology vendors (which can limit flexibility and increase costs). Solutions such as a public key infrastructure should be implemented to ensure the security and authenticity of electronic service delivery. Digital signatures should be incorporated into that strategy. A central data center should be established to host all data computing facilities and databases and software applications for all central government agencies. This requires the shutdown of individual agency data centers in order to consolidate data systems and establish a single integrated management system for IT resources. This integrated model minimizes costs, harmonizes business processes, consolidates infrastructure, and leverages data sharing.

3. Implement Projects in Partnership with the Private Sector

The third key lesson for Group B countries is the role of public–private partnerships in accelerating the development of Digital Governance programs in Korea. Korea uniquely utilized partnerships with the private sector to advance national priorities, including those supporting national economic growth, and to meet development targets. From the very beginning, when Korea made the decision to focus on the ICT sector as a national priority, the government worked closely with the private sector to deliver on its development targets. The government played a strategic role, supporting the development of foundational technologies for e-Governance and working closely with domestic private sector market leaders to make investments in technology, people and projects (see box 6.9).

A related objective of the public–private collaboration was the creation of new private sector jobs in the ICT sector and business growth through the development and implementation of IT projects for the government. The strategy was developed with the objectives of addressing the negative impacts of the downsizing of the labor force, privatization, and restructuring of institutions and processes, while also seeking to improve the government's productivity and enhance the delivery of services to citizens (Song 2007; Special Committee for e-Government 2003). This initiative provided an impetus for retraining staff, both in the private and public sector, to equip them with skills in IT, which, as a sector, grew quite quickly during this period. Training programs were offered by private sector organizations and were provided to over 30,000 public sector staff every year.

A variety of contracting models were considered for development and maintenance of e-Government projects, such as outsourcing and public–private partnerships (PPPs). Outsourcing involved transferring portions of work to external

---

Box 6.9 Public–Private Investments in Technology

An R&D consortium of government-funded research institutes, academia and private companies worked together in the 1980s to develop a lower cost electronic switching device, with a budget of USD 60 million. This was rolled out by Korea Telecom (which was government-owned at the time), allowing Korea to achieve 100% telephone penetration in the country within a short timeframe and with a considerably lower budget than would have been needed if the device had been imported. In the late 1980s, when Korea launched the development of the NBIS, the government joined hands with private companies—Samsung Electronics, LG Electronics, Hyundai Electronics and Trigem Computer—and a government-funded research institute called ETRI to design and develop a domestic mid-size computer, which was then purchased by the government and deployed to central government agencies and local governments. In this way, it achieved two outcomes—personal computer penetration across government laid the foundation for e-Governance in Korea, and the purchase of domestic computers from the Korean private sector boosted the high-technology market.

To build a critical high-speed broadband network in the 1990s, the government put together a public–private working commission, including public agencies—MIC, NIA and ETRI—and private agencies—Korea Telecom and Dacom—to develop a basic implementation plan. The private sector was free to invest in KII-Public, which was intended for households and individual consumers. The government invested USD 1 billion in KII-Government to connect government agencies, raising funds from telecom operator revenues. The rollout was carried out by Korea Telecom and Dacom. Once again, the government played a strategic role, working closely with domestic private sector market leaders to make the necessary investments.

---

suppliers instead of completing them in-house. PPPs involved both private and public sector partners working together to design, plan, construct and operate ICT projects. The private sector also helped the government to develop a common standard for government, called eGovFrame, for an Open Source software development environment.[16] The goal was to provide greater flexibility for smaller vendors to compete for e-Government projects and to minimize vendor lock-in by use of proprietary software and frameworks that tend to provide larger vendors with a competitive advantage. At the project implementation level, a variety of methods were used to contract with the private sector, sometimes unsuccessfully.

Reflecting on Korea's approach, PPPs were critical to the success of Korea's Digital Governance programs. The government worked closely with the private sector to develop high technology components at a lower cost, attract investments into critical broadband infrastructure which eventually enabled service delivery to citizens, and to help create jobs in the midst of a financial crisis. The private sector was also an instrumental partner at the project implementation level as contractors, although some of these arrangements worked less well than others.

Critical Success Factor: A Strong Partnership between the Public and Private Sector

A number of countries, including the United States, make use of PPPs to advance their e-Governance programs. In the US, a privately owned, ICT service provider called NIC has successfully competed to finance, implement and maintain e-Government projects for 28 state governments, in the form of a PPP. NIC recovers the cost of its investments in developing systems by charging fees from businesses (their most lucrative segment) that stand to gain from fast-track government-to-business services. Contract durations may last between 5 and 20 years, during which time, it provides maintenance services for the systems it has built on behalf of the government, ensuring continuity of operations. NIC also hires local staff from states as contractors to enhance ICT-related job opportunities within the state.

PPPs are a critical part of the Korean success story. The Korean government worked with the private sector on R&D to develop critical technologies at a lower cost, domestically. In partnership, they made critical investments in network infrastructure to help stoke the demand for electronic services. These partnerships were also key to creating jobs in the ICT sector. At the project implementation level, the government employed private contractors as part of the contracting model. It contracted with individuals to help deliver services in areas where it had a skills shortage, and engaged private sector vendors to help create common software development standards for a whole-of-government approach to building and managing IT infrastructure. Aspiring countries should build strong links between the public and private sector to enable the success of their e-Governance programs.

Policy Implication for Group B Countries: Bring critical technical and functional skills, structures and services from the private sector into the public sector

Although the technical skills of the private sector are needed by the public sector, it is important that contractors feel ownership for any digital product that they are building for citizens. Creating a winning solution is more than checking boxes on a deliverable schedule. It is creating something usable and useful that can help the institution/project/team reach their goals to make a positive impact in the lives of people and communities. Often, contractors feel detached from the process, allowing them to blame the client for poor direction or management, and allowing the client to blame the contractor for failure. Contracts should be defined to operate as PPPs, so that accountability for delivering public services is shared by the two parties.

PPPs not only enable the transfer of skills, thus boosting investment and facilitating the implementation of digital governance programs. They also generate training and employment opportunities for citizens and public sector staff. In Korea's case, PPPs stimulated IT industry development and generated much-needed jobs in the private sector following the Asian Financial Crisis.

However, the Korea model may not be replicable. Essential prerequisites for a PPP arrangement to work are trust, without which governments will not open up to the private sector, and the existence of a credible private sector.

Learning from Mistakes…How Did Korea Cope with Setbacks?

While we enjoy learning from success, most learning comes from failure. These are often reflected in practices and decisions that could have been anticipated or managed better and, in some cases, avoided. Development of country-wide ICT infrastructure and governance institutions is a complex multi-decade undertaking. It stretches government capacity and resources, and requires flexibility and adaptation to manage the constant changes in technology. Given that this is such an organic and complex process, mistakes will be made. The critical issue is to learn from these mistakes and reverse them quickly, so that, over time, digital governance is increasingly associated with improvement and successful reform, as opposed to inefficiencies, waste, and failure. Even a successful country like Korea made errors in judgement and experienced many problems while planning and developing its digital governance program. These errors generate valuable lessons for other countries embarking on their digital governance initiatives. This section aims to identify the critical problem areas of the Korean experience in order to limit the mistakes of other developing countries on a similar journey.

A close review of chapters 2–5 reveals some of the setbacks that Korea experienced when developing its digital governance program. The lessons learned from these setbacks concerned the critical role of evaluation, empowering project managers to secure funds and make changes when needed, anticipating and managing departmental turf battles and institutional expansion, addressing cross-government inefficiency, avoiding redundancy in ICT investment, ramping up capacity to manage a modern service delivery system, and preparing for and managing systems integration across departments. More specifically, these lessons related to the following areas:

Skills: Many civil servants did not possess the capacity to work under the new public service delivery system. While Korea started out with generalists in the public sector, the government realized that advanced technical specialists in the public sector would be needed to succeed in this field. To address this challenge, technical and functional experts were cross-trained to develop "hybrid" skills that would enable them to design, implement and manage digital governance projects.

Governance Model: The government struggled to find the proper mechanism to manage the cross-government implementation of e-Government. In the early stages, NIA and MIC were given this responsibility, but neither had the power to supervise or manage all the e-government projects of other government ministries and agencies. Budgetary frictions arose between the different government agencies that were competing for limited e-government resources. Moreover, due to the excessive competition to expand the business and civil service boundary of the organization, each ministry over-invested in e-government infrastructure, which then proved to be redundant. This was often exacerbated by multiple competitive procurement practices between government agencies. In order to address the challenge of managing departmental tensions given the limited resources for implementation, Korea experimented with innovative governance models to encourage inter-agency collaboration. Furthermore, an EA frame-work—importantly, mandated by law—helped rationalize duplicate and over-investment in infrastructure, while the use of financial instruments helped prioritize budget allocation and investments.

Service Delivery: Departments often defended their own parochial interests, without adopting a benign service delivery approach. For example, as Korea rolled out its Government Data Center initiative to facilitate data sharing, many departments were conflicted about the initiative, as it required data and resource sharing. Some of these departments responded by restricting access to their databases through both regulation and legislation. The result was that services requiring information from multiple government ministries and agencies (such as passport issuance) were not able to be provided as an online service. These actions undercut the efficiency and value of the proposed solution. In retrospect, antici-pating the mixed incentives of departments to the change would have been key to managing this lack of cohesion.

Partnerships With the Private Sector: Some partnerships with the private sector were less successful than others.

- *Tax Information System.* The Tax Information System (TIS) project procured a consortium of four private contractors to design and construct the system. However, the development took 3 years, due to various miscalculations. The private contractors did not have a lot of experience in developing a system of the scale involved. In addition, the tax authorities faced problems of division of labor and responsibility between the contractors as well as the sequencing of workflow and processes assigned to different contractors. The experience brought home the lesson of devising a clearly defined project scope and closely examining private contractor capacity to implement complex systems (Korea Eximbank 2013).
- *e-Procurement System.* In another example, KONEPS, the e-Procurement system, initially outsourced the maintenance and operations of the system to IT staff from the private sector, as the use of the system had increased to such levels that it was hard for government officials to manage on their own. However, this led to a decline in the professionalism of government employees as reliance on private contractors increased. While the initial contract was for just a year, it was subsequently extended to 5 years, causing frequent changes in the contractors who were assigned to the project, thereby leading to con-cerns related to the lack of continuity and system stability. Moreover, as the private providers were not motivated to improve service quality or delivery, user demands were rarely met promptly or within appropriate service stan-dards, causing complaints from KONEPS users. Learning from the lessons of the initial model, the KONEPS team introduced a long-term outsourcing

contract of 3 years with a pre-arranged service level agreement (SLA), introducing stability and continuity of private contractors, and improvements in service quality. Costing was based on a variable system in which fixed payments were calculated based on additional operations, which improved service quality delivered by the contractors.

Interoperability: There were also lessons related to the development of ICT infrastructure. In the early stages, Korea's software development was based on open source technologies. While this was a good idea, it also required close attention to interoperability. In Korea's case, the melding of systems was mired in compatibility problems and excess costs. In retrospect, early emphasis on common standards and an interoperability framework for whole-of-government would have helped mitigate the downstream costs of rationalizing and harmonizing systems across government.

IT Budgeting: While Korea aimed to manage the program according to a predetermined cost and schedule, at times quality suffered. Project duration was fixed at one year, so any delayed and/or unfinished projects were not allowed to secure the next phase of the project budget, which degraded the quality of the delivered services. Inefficiencies were further evidenced in the sizable number of pilot projects that were abandoned, resulting in wasted public resources. Many of these were linked to hasty predictions of future technology that the market was unprepared to supply. Korea addressed these challenges by moving to a medium- to long-term budget plan of at least five years.

Evaluation: As chapter 5 on Impacts shows, Korea's e-government projects often lacked detailed and careful post-evaluations that are necessary to prevent repetition of the same mistakes. In recent years, Korea has placed a particular emphasis on baselining projects for evaluation, and on data-driven governance to measure results, outcomes and impact. Addressing these issues right from the start, 20–30 years ago, would have provided a rich source of data to evaluate projects, services, and processes, and to draw empirical lessons not only for Korea but for other countries as well. As a result of this learning, Korea now stands in the company of two other OECD countries (Demark and the UK) that are able to report and account for almost all financial benefits realized through ICT projects. Most countries can report and account for no more than 25% of direct financial benefits, which makes it difficult to build a business case for future investments, to get sustainable support and funding, and to make decisions on investing in alternative options (OECD 2015).

Opportunities for Leapfrogging

Korea's experience offers several opportunities to accelerate sustainable development by leapfrogging over more expensive or less efficient technologies, jumping directly to more advanced ones, or helping avoid some of the mistakes that have already been made. Broadband and internet penetration, for instance, have made many more technologies and services available than those that needed to be created from scratch when Korea started its journey. Another major technological change has been the rapid proliferation of the mobile platform. In the African continent, cellular subscriptions have outstripped fixed lines that entail exorbitant overhead costs, opening up a host of opportunities for leapfrogging. For countries seeking to learn from Korea's Digital Governance experience to leapfrog to advanced solutions, two areas in particular may be instructive: (1) Transitioning to the Cloud to maximize the efficiency of shared resources, and (2) Building in Smart Governance for a more data-driven and service-oriented government.

Transitioning to the Cloud: While an integrated government data center is theoretically a foundational element for digital governance, in reality, few early stage countries are able to develop a consolidated government data center. An integrated data center requires cohesion and collaboration between public agencies, which tends to require planning, momentum, and capacity—all of which need time to develop. In Korea's case, the data centers of each government agency were all shut down, while an integrated data center was established. Since this resulted in a large and integrated government data center, provision of cloud services was possible. Cloud computing, or in simpler shorthand just "the cloud," focuses on maximizing the effectiveness of the shared resources. Cloud resources are usually not only shared by multiple users but are also dynamically reallocated to users according to demand. Government-supplied cloud computing services are an advanced form of service. However, such services are only available when the supplying service range is determined and the government is well equipped with a sufficient infrastructure to provide these services. For many early stage countries, borrowing private facilities is an option if such infrastructure is not fully available. It may be more cost-efficient for such countries to utilize private services before such an infrastructure is established. It would also be beneficial because a cloud service can increase the efficiency of governmental duties. Utilizing private services before converting to cloud computing for government is called a transition service. A phased transition service can be a valuable option for early stage countries.

Building in Smart Governance: Most governments commence digital service delivery through Gov 1.0 and Gov 2.0, which emphasize citizen access to services through the internet, as opposed to transacting in person using a combination of manual and automated processes at physical offices. While Gov 1.0 provides one-way transactions, Gov 2.0 provides a virtual platform facilitating participation, two-way discussion and citizen-generated content through advanced e-government technologies, to provide services to citizens, which could potentially lead to changes in political and governance structures (O'Reilly 2010). Gov 3.0 or Smart Governance refers to governments using data for governance, or governments in the age of the semantic web, or government as a social machine in an ecosystem (Berners-Lee and Fischetti 2000). Gov 3.0 entails the delivery of personalized or bespoke services to citizens by building competent and transparent government, innovating administrative practices and processes, and introducing a hyper connected e-government.[17] However, delivering government services over the internet invariably suggests some blind and overlapping spots for services, especially for those who are socially vulnerable and digitally marginalized. In Korea, Gov 3.0 emphasizes O2O (online to offline) services to citizens by way of combining virtual and physical spaces through hyper-connected devices such as IoT (Internet of Things), cloud computing, big data analytics, mobile devices and other intelligent technologies. In Gov 3.0 environments, Korean officials do not wait passively at the office for the digital access and applications of citizens needing administrative services, but actively visit the blind spots armed with sophisticated digital devices and provide services to citizens in need. Early stage countries, grappling with poor connectivity that hinders digital delivery of services to citizens, may consider designing O2O services, particularly through mobile platforms, to bridge the access to services gap for the poor, vulnerable and marginalized.

Conclusion

Whether a country is rich or poor, or large or small, Korea's e-Government experience provides an abundance of lessons for countries pursuing Digital Government reform. We have applied Korea's successes and challenges to two groups of countries—Group A, with an extreme poverty rate above 40% that require support for urgent needs, and Group B, which are low/middle income countries, and have more complex needs.

For Group A countries, the key lessons stress the importance of ensuring sustained, high-level leadership and support for digital governance as a national priority; having staff with hybrid technical/functional skills; improving inter-agency collaboration; and sequencing the development of foundational infrastructure components, pulling together common standards for a whole-of-government approach to benign service delivery.

For Group B countries, in addition to those outlined above, the key lessons include the creation of a local governance model that increases frontline officials' interaction with citizens and heightens their understanding of service needs; integration of applications, services, processes, data, and technology to coordinate seamlessly with peer and subnational agencies; and the establishment of partnerships with the private sector to advance national priorities and meet development targets.

As this chapter demonstrates, not all of Korea's decisions and actions were good ones. In addition to the successful decisions and strategies, we have also tried to highlight some of the setbacks Korea faced so that other countries can anticipate these problems and try to address them early on. These setbacks have pointed to the importance of:

- baselining projects to facilitate monitoring and evaluation,
- monitoring performance implementation problems and making necessary changes when expected cost, deliverables and timelines are missed,
- being purposeful about acquiring hybrid technical/functional skills in government to design and manage these projects and ensuring skills transfer,

- finding an appropriate governance model to embed and manage the digital governance program across government agencies,
- anticipating the varying interests of different agencies and departments to manage changes and improvements in service delivery, and requiring citizen-focus as a means of avoiding self-interest,
- placing an early emphasis on common standards and a whole-of-government approach to service delivery, and
- putting in place medium- to long-term budgets for IT-enabled projects to account for multi-year implementation and foster high quality while still managing to cost and schedule.

Despite the setbacks, Korea's top ranking in the IDI is indisputable and, by any standard, reflects remarkable success. Based on this accomplishment, this paper has aimed to shed light on the country's objectives, strategies and approaches, which have resulted in the creation of a holistic digital governance program that supports efficient resource management and public service for its citizens. We have focused on the critical governance elements of an e-Government initiative as well as the complex technology and infrastructure elements. We have synthesized lessons learned from what worked as well as what did not work. Although we have recommended a framework for sequencing, we have also acknowledged that e-Government, even in the best of circumstances, entails challenging and at times conflicting dynamics.

Above all, throughout this enormous undertaking, Korea has maintained extraordinary commitment (as evidenced by decades of involvement by the political leadership and the participation of a highly digitally-engaged citizenry) and strong resilience (apparent in its never-ending ability to chip away at problems until they were solved). It is with this spirit that we commend Korea and express our gratitude for sharing its experience with countries around the world.

Annex 6A: Guidance for Sequencing a Digital Government Strategy Based on the Korea Experience

	Phase I: Lay the foundation	Phase II: Introduce digital services	Phase III: Build for whole-of government
	Countries in early stages: Group A	*Countries in middle stages: Group A & B*	*Countries in advanced stages: Group B*
Legal/regulatory framework	• Establish laws and regulations to create adequate infrastructure for an e-Government system, with particular emphasis on the creation of high-speed broadband capacity (wired and wireless) throughout the country. • Laws and regulations may pertain to the development of the EA, the establishment of institutions to develop and manage the EA, or other aspects of establishing an adequate infrastructure for an e-Government system.	• Establish laws and regulations to facilitate the creation of digital services. • Laws and regulations may include those related to mobile services, security, digital signature, privacy, and others.	• Develop the legal and regulatory framework to support complete "whole of government" integration. • Establish laws and regulations to support data-driven governance. This entails the development of laws and a regulatory framework which support the use of data to make better policy decisions and to improve the collaboration and trust between citizens and government. • Examples of these types of initiatives that may require changes in the legal framework are open data, knowledge management, customer relations management, performance management, and others.
Bureaucracy	• Commence capacity building to create a "hybrid" set of knowledge and skills in the government. This new combination includes those that are sectoral/functional in nature and those which are IT-related. • Strategies involve training and educating generalists and functional specialists in IT knowledge and skills and, when necessary, recruiting new "hybrid" staff.	• Train, educate and acquire staff with digital services skills and experience. Ensure that relevant officials and staff with "hybrid" skills have specialized digital services expertise. This requires knowledge and skills involving automation of both front- and back-end service delivery.	• Develop skills to promote vertical and horizontal integration of systems across government. These skills tend to be non-technical in nature, involving strategic vision, collaboration, teamwork, communication, and management. These skills are essential for leading and implementing e-Government changes which involve all government agencies and, in particular, require that these agencies share data and systems and, more generally, work together.

table continues next page

| | *Phase I:* **Lay the foundation** | *Phase II:* **Introduce digital services** | *Phase III:* **Build for whole-of government** |
	Countries in early stages: Group A	*Countries in middle stages: Group A & B*	*Countries in advanced stages: Group B*
Organization	• Establish a body at the highest level of government to manage cross-agency coordination. • Establish clear lines of responsibility and accountability for e-Government design and execution at the agency level.	• Use the leadership of the high-level coordination body to ensure that digital governance is used as one of government's primary methods for innovating, serving and improving citizens' access to services and the quality of the delivery of those services, across all government agencies. • Provide leadership and oversight to initiate government-wide automation and digital delivery of services. • Leverage the high-level coordination body to address cross-agency issues which must be resolved to successfully implement digital services.	• Evaluate the effectiveness of the organizations involved in directing and implementing e-Government policy and strategies. If certain organizational approaches are not working, experiment with other approaches. Be open to changing agencies and/or committees leading or involved in the e-Government agenda. Leverage leadership at the highest level to manage the resistance to change which accompanies integration of systems across agencies.
Empowering Local Government with Citizen-Centered Solutions (Performance and Accountability)	• e-Government plans and strategies should have a robust component focused on empowerment of local government to deliver services to citizens in a more efficient, user-friendly manner. • Local government needs to be a critical partner in the e-Government design and implementation process.	• Ensure that local government is a key player in digital services. • Empower local government to initiate digitization of services for all local services. • Digital service delivery strategies should be designed with input from citizens, and should be tailored to reduce cost, time waiting for processing, and number of visits required to secure a service.	• Mechanisms for securing citizen feedback and enhancing citizen engagement should be deepened and expanded by service area, supporting the development of a strong evidence-based system of citizen feedback by service. • Ensure citizen feedback and engagement is linked to the delivery of more complex services, such as those involving multiple agencies. • Integrate data related to citizen feedback across agencies in order to compare performance of service delivery across government.

table continues next page

	Phase I: Lay the foundation	Phase II: Introduce digital services	Phase III: Build for whole-of government
	Countries in early stages: Group A	Countries in middle stages: Group A & B	Countries in advanced stages: Group B
Architecture/Infrastructure	• *Technology Plan and Approach:* Develop an Architecture Plan which provides (a) a clear sequencing of the key building blocks of the infrastructure, and (b) standards for government agencies in building e-Government systems. These elements are critical for scaling up e-Government across government. Importantly, the creation of adequate high speed broadband capacity (wired or wireless) across the country to serve government agencies (public), citizens and businesses (private), and research centers/academia is foundational. • *Enterprise Architecture:* Create the EA required to establish a foundation for intra- and inter-governmental connectivity and ICT resource management. • *Open Source Framework:* Create an Open Source framework for developing software required for e-Government systems. The Open Source approach avoids dependence on private technology vendors (which can limit flexibility and increase costs). An emphasis on standardization is key in order to minimize costs associated with integration of software, processes, and systems across agencies.	• *Front- and Back-end Development for Streamlined, Citizen-centered Service Delivery:* Develop front- and back-end systems to support a citizen-centered service delivery system—focusing on reducing citizen time, cost and number of visits required to secure a service. • Integrate citizen feedback and grievance redress mechanisms for each service into the system design. • Track service delivery progress and capture data for delivery improvement for each service. • Leverage mobile devices for data collection, data storage, notifications and tracking, citizen feedback, and exchanging photos and video if applicable. • *Citizen-centered Process Re-engineering:* Reengineer processes with the convenience of citizens in mind, so that citizens will not have to interface with multiple agencies in order to receive information or services or to interact with the government. The goal is for government agencies to work together seamlessly to deliver services to citizens through a number of channels—online, government office counter, call centers, and so forth.	• *Horizontal and Vertical Systems Integration, with Citizens as the Focal Point:* Integrate vertical and horizontal systems across government to establish a whole-of-government ICT infrastructure which emphasizes interoperability, compatibility, elimination of duplication and redundancies. In addition, the integration should have a citizen-centered focus. • *Integration of Systems through Priorities and Projects:* Establish projects and priorities to promote horizontal integration of systems. Examples of systems integration projects include automated document management, integrated financial management, local government information management, human resources information management, integrated procurement management, and others. • *"Data-driven Governance."* Establish a government culture of "Data-driven governance," which means that governments use data and analytics from transactions to make decisions or policies.

table continues next page

Phase I: Lay the foundation	Phase II: Introduce digital services	Phase III: Build for whole-of government
Countries in early stages: Group A	*Countries in middle stages: Group A & B*	*Countries in advanced stages: Group B*
	• *Security and complete automation of digital services.* Implement solutions to ensure security and authenticity of electronic service delivery. Incorporate digital signatures. • *Integrated Government Data Center.* Establish a central data center to host all data computing facilities and databases and software applications for all central government agencies. This requires the shutdown of individual agency data centers in order to consolidate data systems and establish a single integrated management system for IT resources. This integrated model minimizes costs, harmonizes business processes, consolidates infrastructure, and leverages data sharing.	

Annex 6B: Enterprise Architecture and Common Standards for Government in Korea

An EA[18] reference model was drawn up for the whole of government in 2004, covering the 31 e-Government priorities. The Korean EA is called ITA or Information Technology Architecture, comprising a business reference model, technology reference model, service reference model, and a data reference model. A 2005 law made ITA mandatory for all agencies for a more integrated government. By October 2012, 15,000 e-government systems belonging to 1,400 public institutions had been integrated into the EA, rationalizing and harmonizing investments in software and hardware across all agencies. Business process reengineering was undertaken with the notion of vertical integration—where local systems are linked to similar higher level systems with similar functionalities—and horizontal integration—where systems are integrated across disparate functions (Choi 2013a; Layne and Lee 2001).

From a **process integration** perspective, a number of processes were reengineered with the convenience of citizens in mind, so that citizens would not have to interface with multiple agencies in order to receive information or to interact with the government. The goal was for government agencies to work together seamlessly to deliver services to citizens through a number of channels—online, government office counter, call centers, and so forth. For instance, prior to the G4C system, a citizen who moved from one jurisdiction to another would have to visit multiple administrative offices in person and make separate civil applications for a transfer, car change, school transfer, and business registration. A redesign of the business process was carried out, utilizing feedback from citizens to improve the civil applications process, using the OPEN[19] application. It was integrated into a single form with appropriate data elements routed to the car registration system, the education information system, tax system and the citizen identity management system, respectively (Choi 2013a).

From a **service integration** perspective, agencies that were previously operating as stand-alone agencies integrated with other agencies to present a whole-of-government interface to citizens. The SSIS provides a good example of such improvements. The introduction of a Unique Resident Number to each individual based on his or her birth registration, required the linkage and integration of different databases and services (Korea Eximbank 2013). The SSIS system is also an example of standardization of information (such as that related to income and property), databases and services across national and subnational agencies. The system allows for the selection of beneficiaries and facilitates decisions on individuals and household benefits. Prior to integration and standardization efforts, different programs used different criteria, operational definitions, measurement and data-points to screen eligible beneficiaries. SSIS brought together various agencies responsible for social welfare programs, and came to an agreement on standard methods and definitions for measurement and screening. 37 different application forms and other documents were unified into a single form (with 5 supplementary forms) that could be used to apply for 100 social welfare programs, addressing the issues of duplicate information, redundancies and the principle of not asking a citizen for the same information more than once (Korea Eximbank 2013).

From a **data and technology integration** perspective, EA standards were applied across all government agencies to consolidate infrastructure that was previously managed separately by individual agencies. A government integrated data center (GIDC) was established to host all government data computing facilities, databases and software applications. Datasets, services and processes were integrated onto a Public Cloud computing platform, called G-cloud, developed by the government.[20] Government agencies were asked to shut down their individual data centers and to consolidate data centers within the GIDC, which would provide an integrated management system for IT resources. Agencies that were prone to restricting access to databases by using legislative and regulatory maneuvering, were asked to share data across government to minimize redundant investments and to facilitate an integrated approach to delivering e-Government services. Over 50% of government computing services were targeted to be moved to the G-cloud. Furthermore, the government led the implementation of PKI (public key infrastructure) as an element of the software infrastructure. To support the use of e-government services safely, the implementation of a PKI became the policy that the government fully supported. Additionally, for citizens, an identity verification system was established (based on the i-Pin, a registration number to receive online services in Korea) to check cyber-crime and cyber-bullying while protecting individual freedom of speech and privacy.[21]

In terms of building **common standards**, the Government made a strategic choice to develop systems using an open source approach. The purpose was to eschew proprietary systems to minimize government dependence on private software technology vendors who each have their own framework for software development. Software development frameworks developed by private vendors are usually a black-box so as to maintain a competitive advantage over other vendors and external parties, as only those who know the framework can maintain the system (Kim and Teo 2013). Initially, the open source approach did not have a standardized framework for government, which resulted in substantial costs with regard to integration of software, processes and systems across agencies. To mitigate these risks and to protect government investments, NIA and MOI, developed an eGovFrame, which included a standardized set of software tools and a library of reusable OSS e-Government components for application development, integration, maintenance and reuse of applications. The e-GovFrame was launched in 2008. It became mandatory for vendors applying to develop e-Government applications through a Request for Proposal (RFP) process to use the e-GovFrame for software development. By 2014, more than 350,000 developers had used the e-GovFrame for 450 projects across government agencies, with a budget of USD 1.26 billion, weakening vendor lock-in and adopting a flexible framework.[22] While previously, 80% of government projects would be awarded to large vendors, more than 60% of projects are now awarded to small vendors, enhancing the competitiveness of smaller vendors in implementing e-Government (Kim and Teo 2013).

Notes

1. The IDI is based on 11 indicators, clustered by access, use and skills. Access includes infrastructure and access indicators such as 'percentage of households with internet access', 'mobile-cellular telephone subscriptions/100 inhabitants' etc. Use includes usage indicators such as 'percentage of individuals using the internet'. Skills include proxy indicators, such as 'adult literacy rate', 'gross enrollment ratio secondary level' etc. Source: ITU, 2015. http://www.itu.int/net4/ITU-D/idi/2015/#KOR

2. Home Tax Service system—See chapter 4.

3. The IT New Deal Projects were designed to hire, train, and help citizens gain IT skills and find jobs servicing government IT contracts.

4. The World Development Report 2016 refers to these aspects as the digital and 'analog' complements.

5. In contrast, Singapore, a top performer on the UN e-government readiness index, which sustained its investments in technology to improve governance outcomes, has been a one-party state since 1959.

6. Presentation at World Bank by Joon-Kyung Kim, KDI President, Jan 11, 2016.

7. The postwar era saw a highly mobile and motivated Korean public who aspired to better themselves personally and nationally. A frequently heard phrase in Korea is 'palli palli', meaning 'hurry-hurry', reflecting the desire for action and agility.

8. One of the remedies, borrowed from the business world, has been to reconcile systems across agencies through an enterprise architecture and interoperability approach, integrating data horizontally across peer-level agencies for cross-cutting functions such as disaster management or homeland security.

9. MOGAHA was later transformed into the Ministry of Public Administration and Security (MOPAS), and then the Ministry of Interior (MOI). The Commerce, Industry and Energy Ministry was transformed into the Ministry of Knowledge Economy (MOKE).

10. Business Process Reengineering follows principles from the work of Hammer and Champy (1991), who argue that the benefits of digitization (including e-government) will only emerge if digitization is twinned with process simplification and redesign.

11. The DBAS/dBrain (IFMIS solution of the Ministry of Strategy and Finance of the Republic of Korea) is the winner (first place in Category 4, EAP region) of the 2012 United Nations Public Service Awards (UNPSA) for promoting a whole-of-government approach (http://english.mosf.go.kr/).

12. Information provided in a brochure on Union Digital Centres, distributed by Access to Information (A2I), Prime Minister's Office, Bangladesh. http://www.a2i.pmo.gov.bd.

13. Former Washington DC CTO, Susan Peck said in 2005 in an interview with Jeanne Ross of MIT Sloan School, "*As a District, the finest thing I can do for you, residents, is to give you benign service delivery. I can make it easy for you to deal with me. I can make it not horrible.*" (Ross et al. 2006).

14. Author's discussions with Prof. Hee Joon Song, Chairman of Prime Minister's Gov 3.0 Committee.

15. An example is the Seoul Night Bus in Korea. Data showed that low-income commuters needed a low-cost option between midnight and 5:00 am to save on taxis. Nine routes were set up based on cellphone calls and texting data (Sung and Rios, 2015).

Bringing Government into the 21st Century • http://dx.doi.org/10.1596/978-1-4648-0881-4

16. Open-source software (OSS) is computer software with its source code made available with a license, in which the copyright holder provides the right to study, change, and distribute the software to anyone and for any purpose. Open-source software may be developed in a collaborative public manner.

17. Author's discussions with Prof. Hee Joon Song, Chairman of Prime Minister's Gov 3.0 Committee.

18. Enterprise Architecture (EA) "is the organizing logic for business processes and IT infrastructure, reflecting the integration and standardization requirements of a company's operating model" (Ross et al., 2006). It can be considered a blueprint reference or guide for organizations, much like an architectural blueprint provides a design for constructing a building. If many users use services generated from a vast and complicated system, then applying EA to the system would be much more effective.

19. Citizen comments and feedback are solicited both openly, for all to see, and individually, to receive a response to their requests or questions. The turnaround time for replies from the government is 3 working days.

20. Cloud computing and storage solutions provide users and enterprises with various capabilities to store and process their data in third-party data centers. It relies on sharing of resources to achieve coherence and economies of scale, similar to a utility (like the electricity grid) over a network. At the foundation of cloud computing is the broader concept of converged infrastructure and shared services.

21. Understanding Korea's identification system (Lee, 2009).

22. Vendor lock-in involves the development of proprietary software with one vendor and, ultimately, being forced to upgrade with the same vendor. This type of lock-in relationship limits the options of a government as its IT systems evolve over time. Vender lock-in may have significant cost implications. In order to avoid vendor lock-in, Korea used an open source approach to software development.

Bibliography

Berners-Lee, T. and M. Fischetti. 2000. *Weaving the Web: The Original Design and Ultimate Destiny of the World Wide Web by Its Inventor*. HarperInformation.

Bretschneider, S., J. Gant, S. Kim, H. Choi, H. Kim, M. Ahn, and J. Lee. 2005. "E-government in Gangnam District: Evaluating Critical Success Factors." Center for Technology and Information Policy, Maxwell School, Syracuse University, Syracuse, NY (project report submitted to Gangnam-gu).

Bussell, J. 2012. *Corruption and Reform in India: Public Services in the Digital Age*. Cambridge University Press.

Choi, H., and S. Hahn. 2008. "Interaction of Information Technology and Organizational Restructuring Strategies: Cases of TIS and HTS in the National Tax Service." *Korean Public Administration Review* 42 (1): 323–44.

Choi, C. 2013a. *Innovative Governance through E-Government*. E-Learning Module. World Bank, KDI School of Public Policy and Management.

Choi, J-W. 2013b. "Good Governance and E-Government." Korea University.

Dunleavy, P., H. Margetts, S. Bastow, and J. Tinkler. 2006. "New Public Management Is Dead—Long Live Digital-Era Governance." *Journal of Public Administration Research and Theory* 16 (3): 467–94.

Eom, S., and B. Kim. 2005. " An Empirical Study on the Influence of Public Infomatization on the Number of Employees and Middle Management in Korean Local Governments." *Korean Journal of Public Administration* 14 (3): 155–84.

Fountain, J. E. 2001a. *Building the Virtual State: Information Technology and Institutional Change.* 61–82.

———. 2001b. "The Virtual State: Transforming American Government?" *National Civic Review* 90 (3): 241–52.

———. 2009. *Bureaucratic Reform and e-Government in the United States: An Institutional Perspective.* New York, NY: Routledge.

Hammer, M., and Champy, J. 1993. *Reengineering the Corporation: A Manifesto for Business Revolution.*

Han, S. 2005. "An Empirical Study on the Perception of the Information Technology Investment Effects in the Public Sector." *Korean Public Administration Review* 39 (1): 237–59.

Heeks, R. 2006. "Understanding and Measuring eGovernment: International Benchmarking Studies." Paper presented at the Participation and E-Government: Understanding the Present and Creating the Future, Budapest, Hungary, July 27–28.

Heeks, R., and S. Bailur. 2007. "Analyzing e-Government Research: Perspectives, Philosophies, Theories, Methods, and Practice." *Government Information Quarterly* 24 (2): 243–265.

Im, T. 2011. "Information Technology and Organizational Morphology: The Case of the Korean Central Government." *Public Administration Review* 66 (1): 435–43.

Jung, W., and N. Son. 2007. "A Study on the Performance Evaluation of the Information Network Village." *Journal of Korean Association for Regional Information Society* 10 (3): 19–43.

Karippacheril, T. G., & Tavoulareas, E. 2014. "Getting Digital Service Delivery Right." http://blogs.worldbank.org/governance/getting-digital-right.

Kim, S., H. Kim, and H. Lee. 2009. "An Institutional Analysis of an E-government System for Anti-Corruption: The Case of OPEN" *Government Information Quarterly* 26: 42–50.

Kim, H. J., J. Lee, and S. Kim. 2009. *Linking Local e-Government Development Stages to Collaboration Strategy.* Social and Organizational Developments through Emerging E-Government Applications: New Principles and Concepts: New Principles and Concepts, 275.

Kim, S. 2008. "Local Electronic Government Leadership and Innovation: South Korean Experience." *Asia Pacific Journal of Public Administration* 30 (2): 165–92.

Kim, S., and J. Lee. 2012. "E-Participation, Transparency, and Trust in Local Government." *Public Administration Review* 72 (6): 819–28.

Kim, S. L. and T.S. Teo. 2013. Lessons for Software Development Ecosystems: South Korea's e-Government Open Source Initiative. *MIS Quarterly Executive* 12 (2): 93–108.

Korea Eximbank. 2013. "Supporting Public Management through e-Government Capacity Development." Knowledge Sharing Series: Joint Consulting with MDBs. Sponsored by Ministry of Strategy and Finance, Asian Development Bank, Korea Eximbank, and Sungkyunkwan University.

Layne, K. and J. Lee. 2001. "Developing Fully Functional E-Government: A Four Stage Model." *Government Information Quarterly* 18 (2): 122–36.

Lee, S., and B. Min. 2002. "The Effects of Adopting the Real Name System for Citizen Participation in Websites of Local Governments." *Korean Public Administration Review* 36 (2): 205–29.

Lee, B. G. 2009. *Understanding Korea's "Identity Verification System."* Korea Communications Commission.

Lee, S. 2011. "A Study on Relations between Trust in E-Government and Trust in Government: Focused on the Factors of Truster and Trustee." *Informatization Policy* 18 (2): 49–71.

Mok, J., S. Myeong, and T. Yun. 2002. "Reduction of Administrative Corruption by E-Government: Focusing on Administrative Red-Tapes and Information-Communication Technology." *Informatization Policy* 9 (3): 3–17.

Moon, M. J. 2014. *Informatization Policy and Development: What Does Korean Experience Tell Us?* Presentation. Department of Public Administration. Yonsei University.

Myeong, S., and Y. Choi. 2010. "Effects of Information Technology on Policy Decision-Making Processes: Some Evidences Beyond Rhetoric." *Administration & Society* 42 (4): 441–59.

No, C. H. and C. H. Ro. 1993. *Public Administration and the Korean Transformation: Concepts, Policies, and Value Conflicts.* Kumarian Pr.

OECD. 2015. *Digital Government Performance.* OECD Publishing.

O'Reilly, T. 2010. *Chapter 2. Government as a Platform.* http://chimera.labs.oreilly.com /books/1234000000774/ch02.html.

Pink, D. H. 2011. *Drive: The Surprising Truth about What Motivates Us.* Penguin.

Ross, J. W., Weill, P., and Robertson, D. 2006. *Enterprise Architecture as Strategy: Creating a Foundation for Business Execution.* Harvard Business Press.

Song, H. 2007. "E-Government of Korea: Achievments & Tasks." Informatization Policy, Ewha Womens University.

Song, H. J., and C. H. Oh. 2012. "Knowledge Sharing Series." Asian and Pacific Training Center for Information and Communication Technology for Development.

Special Committee for E-government. 2003. *White Paper on E-government (in Korean).* Seoul.

Sung, N. M., & Rios, M. 2015. What Does Big Data Have to do With an Owl? http://blogs .worldbank.org/transport/what-does-big-data-have-do-owl.

United Nations E-Government Survey. 2014. "United Nations Department for Economic and Social Affairs. E-Government For The Future We Want." http://unpan3.un.org /egovkb/.

Environmental Benefits Statement